Building with
Virtual LEGO

Building with Virtual LEGO: Getting Started with LEGO Digital Designer, LDraw™, and Mecabricks

John Baichtal

New York   Chicago   San Francisco   Athens   London   Madrid
Mexico City   Milan   New Delhi   Singapore   Sydney   Toronto

Cataloging-in-Publication Data is on file with the Library of Congress

McGraw-Hill Education books are available at special quantity discounts to use as premiums and sales promotions or for use in corporate training programs. To contact a representative, please visit the Contact Us page at www.mhprofessional.com.

Building with Virtual LEGO: Getting Started with LEGO Digital Designer, LDraw™, and Mecabricks

1 2 3 4 5 6 7 8 9   LWI   21 20 19 18 17 16

ISBN   978-1-259-86183-3
MHID   1-259-86183-X

This book is printed on acid-free paper.

Sponsoring Editor
Michael McCabe

Editorial Supervisor
Donna M. Martone

Production Supervisor
Pamela A. Pelton

Acquisitions Coordinator
Lauren Rogers

Project Manager
Patricia Wallenburg, TypeWriting

Copy Editor
James Madru

Proofreader
Claire Splan

Indexer
Claire Splan

Art Director, Cover
Jeff Weeks

Composition
TypeWriting

About the Author

John Baichtal has written or edited over a dozen books, including the award-winning *Cult of Lego* (No Starch Press, 2011), *LEGO hacker bible Make: LEGO and Arduino Projects* (Maker Media, 2012) with Adam Wolf and Matthew Beckler, *Maker Pro* (Maker Media, 2014), and *Hacking Your LEGO Mindstorms EV3 Kit* (Que, 2015). He's hard at work on his latest project, a compilation of LED projects for No Starch Press. John lives in Minneapolis with his wife and three children.

Contents-at-a-Glance

Contents

Acknowledgments

Thanks to Matt Wagner for helping make this happen. I also want to thank everyone at McGraw-Hill for believing in the project. My mom, Barbara Baichtal, helped a lot, as did the builders who let me publish screenshots of their models. Finally, I couldn't write a word without my wife Elise and kids Arden, Rose, and Jack.

Building with Virtual LEGO

CHAPTER 1

Building with Virtual LEGO

I HEAR THE COMPLAINT all the time—LEGO bricks, while amazing, just cost too much. A typical price is 10 cents a brick if you purchase them in sets, more if you buy them individually from a LEGO store. It doesn't help things that those bricks are likely to last 20 years or more. LEGO's legendary quality derives from really great plastic that holds its shape and color long after inferior formulations have gone into the trash bin. Called ABS, this plastic unsurprisingly costs a lot to manufacture, and that price is passed on to the customer. In the end it's simple economics: if you want a $200 set and can't afford it, you can't buy it and can't build the model.

Enter the concept of building with *virtual LEGO*. Imagine a design application that allows you to build any model you can imagine on your computer using nearly every brick the company has molded in a vast array of colors. You can see an example in Figure 1-1: any brick in any color! You manipulate the virtual bricks with your mouse to build anything you could build with real LEGO.

In this chapter you'll learn about the three most popular LEGO building programs: LEGO Digital Designer (henceforth LDD), Mecabricks, and LDraw, as well as discovering reasons why you'd want to—or might not want to—use such an application.

What's Virtual Building All About?

With virtual building, your ability to build is not limited by economics, nor storage space in your house, nor even what parts and colors the company currently offers. Of course, there are downsides. You do not end up with a physical model the way you would with the normal way of building. Furthermore, there is no tactile experience of running your hands through a box of parts looking for that one perfect 2 × 3 dark red brick that you need for your latest project. It's kind of like the comparison between buying a book online versus rummaging through a dusty old bookstore hoping to find a gem that you never knew existed. Building physically and building virtually both are great, and both are needed, but it's not an either-or.

The following are some features common to virtual LEGO building programs.

Virtual Building

You can build virtual LEGO models using these programs, just as you could build LEGO models on your dining room table. It's the most obvious aspect, perhaps, but it bears mentioning. You can build anything you could build in real life—and a lot besides. In general, the programs I describe in this book are intended to closely

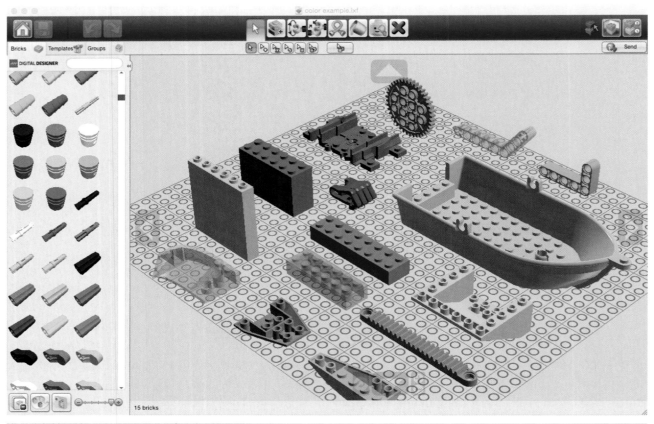

Figure 1-1   LEGO Digital Designer (LDD) offers every part in every color. The rest is up to you!

simulate a physical building experience. The bricks attach together the way you'd expect. Figure 1-2 shows a simplified virtual model that illustrates just how easy it can be.

Robust Brick Palettes

The parts libraries, sometimes called the *brick palettes*, seem to include every imaginable part in every imaginable color. I say "seem to" because it's likely not all of them, but close enough not to matter to most people. Figure 1-3 shows one of the brick palettes in LDD. One thing you will never lack for in a virtual build is finding the right parts or enough of them.

Physics

The virtual bricks you see on the screen aren't merely shapes—they also have the physics of the real brick associated with them. For instance, you can't put a Technic peg into the side of a brick (Figure 1-4 shows me trying!) nor put the small-sized tire on a big wheel. If you rotate an axle in LDD, anything geared to that axle must turn as well. For free programs, these applications do a decent job of simulating the way bricks actually fit together.

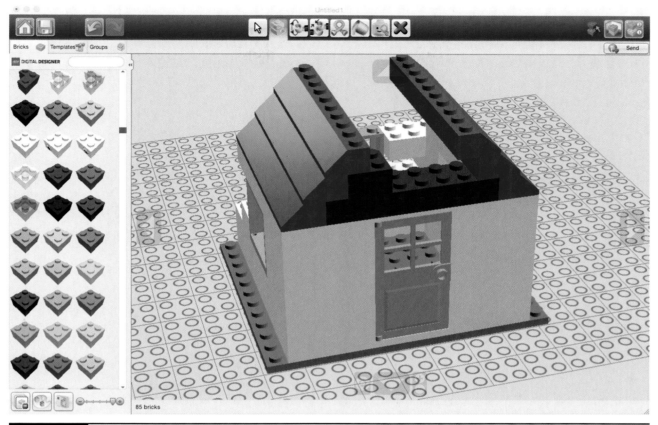

Figure 1-2 A simple model demonstrates what LDD is all about.

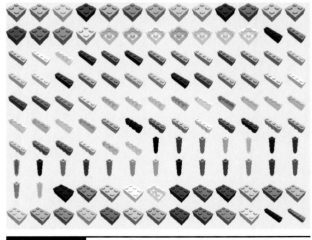

Figure 1-3 One of the brick palettes in LDD.

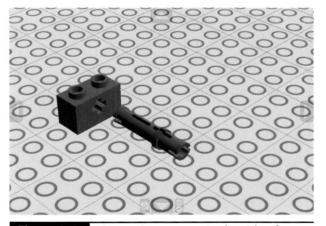

Figure 1-4 You can't put a peg in the side of a brick.

Sharing

The program is optimized for sharing your models with friends, either the actual LDD file itself or outputting web pages with auto-generated step-by-step guides (like the one in Figure 1-5) to building the model. Think about it this way: what use is being a great LDD builder if no one ever sees what you've built? Chapter 9 describes a number of options for creating these guides.

Is Virtual Building for You?

Let's go into depth on eight reasons why this might be a wonderful tool for any LEGO fan. Then, in the interest of fairness, I'll detail three very real reasons why they might not be right for you.

Reasons to Build Virtually

As mentioned earlier, there is a plethora of very cool reasons to use LDD.

1. It's Free

I'm pretty sure I already mentioned that these programs are free, but it bears mentioning again: *they're free!* You never have to buy anything, pay for a membership, buy access to parts, anything! Because LEGO is often criticized for costing a lot, it's good to know that part of the experience won't cost you a penny.

2. It's Easy

Young children and technologically challenged adults can both use these programs. They are something you can play with very readily

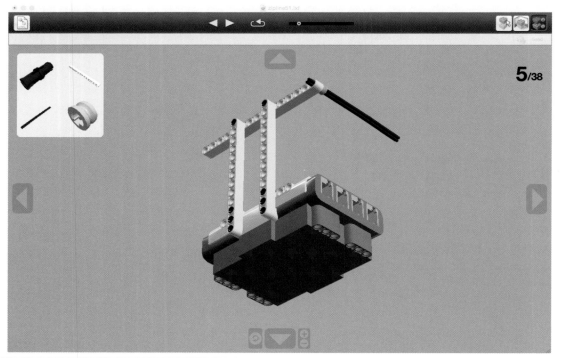

Figure 1-5   Building instructions can be generated with the press of a button.

without reading through a bunch of forums or being forced to learn key commands—though still buy this book, please! The basic mechanic involves selecting parts from a brick palette and attaching them together much as you would a real set of bricks. Like I said, it's very easy to use.

3. Use Bricks That Don't Exist

These programs depict every brick ever created by LEGO in nearly every color the company has employed. But this does not mean that LEGO has actually made these bricks! Figure 1-1 shows some crazy combinations of parts that the company likely will never offer. Building physically limits you to bricks that exist and you own. Virtual building offers no such limitation!

4. Build Epic Creations

If price is no object, how big would you build? LEGO design programs support some seriously huge models—in fact, for LDD, there is no size limit of your models other than the limitation of your computer's memory. If you were to build those same huge projects in real life, the LEGO bricks might cost tens of thousands of dollars. Don't be stymied by humdrum real-world limitations. Go crazy and build big on your computer!

5. Try Before You Buy

Let's suppose that you do want to build a project using physical LEGO bricks. Building virtually first offers the perfect opportunity to create your model for free by allowing you to first build it virtually. The program then outputs a parts list, giving you a list of all the parts you used, without wasting money and effort hunting after bricks you *might* need. You need only buy what you need.

The same goes if you're buying one of those awesome but pricy LEGO sets. You reference the inventory of bricks from that expensive set, allowing you to know exactly what you can do with those parts, before you spend the money. With LDD, you can remix that model to your heart's content before spending a penny!

6. Prototyping

Forget about building a castle or a space ship. What about building a new invention? Wait, build a serious creation in LEGO first? You'd better believe it. Engineers around the world reach for LEGO as the first step toward creating a mechanism. Many tinkerers, hackers, makers, and inventors use LEGO elements to build enclosures, chassis, mounts, and so on. It's simply easier to use a building set to create a mechanical prototype than it is to machine all the parts first.

The PancakeBot (Figure 1-6) is a pancake-making robot that is able to create virtually any shape of pancake. You guessed it, its creator, Miguel Valenzuela, first prototyped it in LEGOs with the design refined in LDD. He created it for his daughters, Lily and Maia, after they inspired him to build a robot that could make breakfast.

Since then, Miguel has showed it off at fairs and conventions—and even the White House science fair. He is selling a professional PancakeBot with a metal chassis, but who could say whether it would have happened without the model first being built on the computer? Learn more about this project at pancakebot.com.

7. Saving Complicated Builds

The next reason to use LDD that I want to mention is that these programs can serve as a digital record of a build simply to help you remember how you did it. Imagine building that

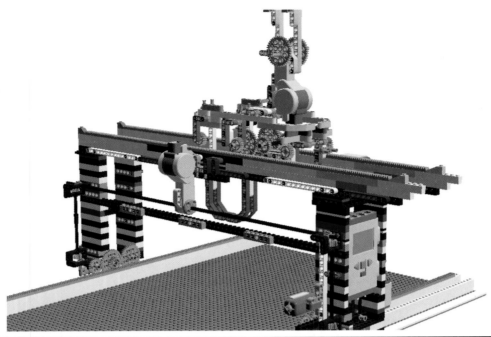

Figure 1-6   PancakeBot's gantry prototyped in LDD. (*Source:* Miguel Valenzuela.)

perfect house, crane, or bridge—if you need the parts for another project, by all means take it apart. But first build it online so that you can re-create it whenever you want. I built the gripper pictured in Figure 1-7, and I know it will come in handy. After I built it in the real world, I replicated it in LDD so that I can reuse the parts for other models.

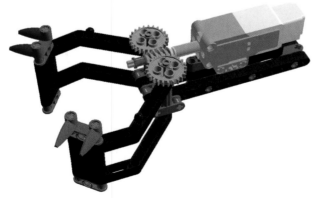

Figure 1-7   I don't need to puzzle out how to build this gripper again because I have it saved in LDD.

8. Sharing

The final and perhaps best reason to use LDD is that it supports the community by allowing you to export renderings, parts lists, and building instructions so that anyone can build your model. I will talk more about this topic in Chapter 9, but suffice it to say that sending your model virtually costs a lot less and involves fewer breakages than sending someone a real model. While this may seem a rather obvious point, it bears mentioning because it underscores just how easy it is to build and share your hobby virtually.

Reasons Why Not to Build Virtually

I hate to bring you down after all those great reasons to use the software, but here are some reasons why you might want to look elsewhere.

1. Compatibility Issues

If you run a Linux-based computer, you'll find that LDD won't work on your system without the help of an emulator. The LDraw Project, by contrast, offers a number of Linux-based alternatives. Mecabricks seems to offer a safe alternative by being a web app, but there can sometimes be compatibility issues with them as well. Basically, what I'm saying is that anytime computers are involved, and especially if you're on a relatively obscure operating system, there will be problems. No one had any compatibility problems when attaching two physical bricks together!

2. Not Professional

There's a downside to free—sometimes you get what you pay for. LDraw and Mecabricks were coded by amateurs, fans who charge nothing to use their creation. Similarly, the LEGO Group offers LDD for free, even though designing and hosting it likely cost them a lot of money.

If you're used to a professional computer-aided design (CAD) program that costs hundreds of dollars to license, you may not appreciate how stripped down these programs are by comparison.

3. No Actual Model

Virtual models are all well and good, but it's not the same as building for real! LEGO had a service a few years ago where you could upload your parts list, and the company would send you the bricks. However, after making a go of it, the company closed down the service, and it's no longer offered.

Overview of Virtual Building Options

I've mentioned LDD, LDraw, and Mecabricks. It turns out these are three very radically different approaches to virtual building. Before we go any further, let's get familiarized with our options.

LEGO Digital Designer

LDD (Figure 1-8) comes from the mother lode: the LEGO Group. It's as official as it gets, which can be good and bad. For instance, being an official LEGO product means that LDD gets cool new bricks added periodically, the new parts having been designed to go with physical LEGO sets. The downside of being a giant company's product is that sometimes those updates are few and far between. The application is freely downloaded from LEGO.com.

An Important Note

Many LDD fans were alarmed to hear first rumors and then a semiofficial announcement that no additional work would be done on LDD. It was, alarmists opined, being canceled altogether.

There was some truth to the rumors, but mostly it's alright news: LDD isn't going away and may still be downloaded from Lego.com. Best of all, the company will still update the brick libraries with new parts as they are released and also release patches for the software.

Now for the bad news: The company isn't looking to create a new version of the program, and no new features are planned. This is only bad news for superfans because the application is already robust and compares favorably with the competition.

LDraw

As I mentioned earlier, each of the three options offers a different building experience. With LDraw (Figure 1-9), it's the focus. The primary purpose of the LDraw organization (LDraw.org) involves the creation and maintenance of the brick library, the all-important palette of parts with which you build your virtual models. Separate from the library, but allied with it, are the bevy of building programs created to

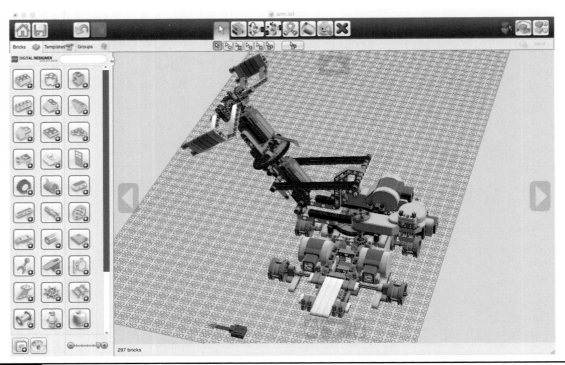

Figure 1-8   Building in LDD.

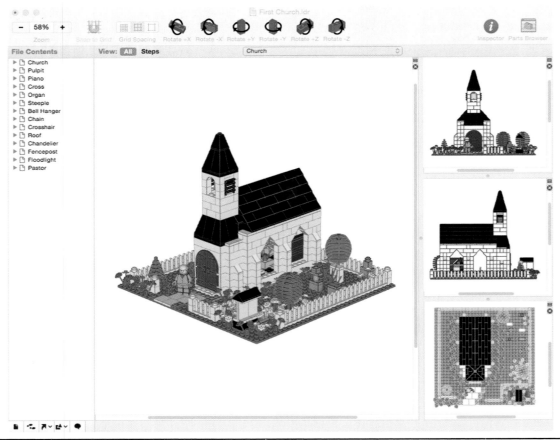

Figure 1-9   The LDraw library is designed to support a variety of LEGO building programs.

manipulate those parts—the CAD programs. No matter what happens to those design programs, their success or failure won't affect the LDraw library.

Mecabricks

Designed by a programmer from New Zealand, Mecabricks (Figure 1-10) makes use of your web browser to interface with an online building program. Don't worry about downloading anything because your creation is always online and always saved. Mecabricks doesn't have a lot of bells and whistles because there's really only one person working on it, but you can still make practically any model you can think of.

Does this seem excessively brief? Don't worry! In Chapter 3, I'll go into detail about what you need to know to choose among the various applications.

Summary

LDD offers many compelling reasons to give it a try. I have listed them out, but now let's explore them! First, however, I want to whet your appetite for cool LDD models. Chapter 2 shows off what is possible to do with the software.

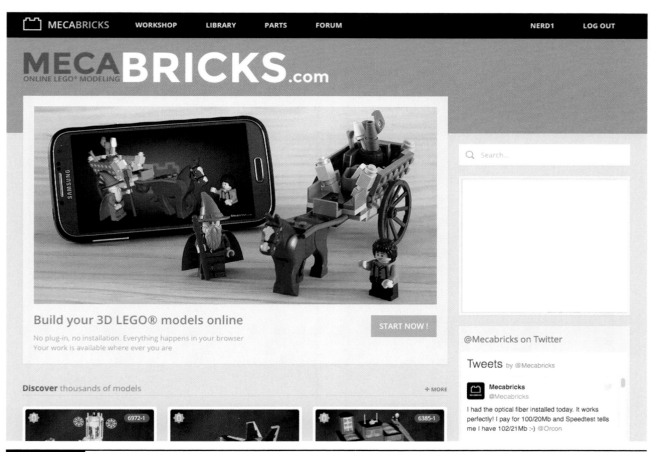

Figure 1-10   Make LEGO models online with Mecabricks.

Gallery of Cool LDD Models

IN CHAPTER 1, I LAID OUT the realities of the LEGO Digital Designer (LDD) program. But what about the possibilities? This chapter details a variety of models, created by myself and other builders, that hopefully will give you an idea of the breadth of ideas possible with virtual LEGO building, ranging from a musical instrument to a scale model of a fantasy temple.

Keytar

This Keytar (Figure 2-1) was built to host a digital synthesizer as part of my book *Make: LEGO and Arduino Projects* (Make, 2012), co-written with Adam Wolf and Matthew Beckler. The Keytar featured a built-in speaker, five octave-controlling buttons, and the twin whammy bars controlling pitch and tone.

I attached Technic bricks to a frame of beams and then placed 8 × 8 System plates on the bricks' exposed studs. The neck was a series of frame bricks with more plates reinforcing it. Built first in the real world and then converted to LDD, the Keytar was something of a design challenge because the model was so huge (3 ft.), and I had never worked on a model that big—either virtually or in the real world!

Great Temple of Odan

You want to talk a big model? Try Mike Doyle's Great Temple of Odan. Figure 2-2 shows a relatively modest fragment of the overall build—only 10 percent of the finished creation's eventual 200,000 bricks.

Mike began building in LDD first because he was able to build as lavishly as he wanted without being limited by real-world considerations. Mike is also known for his beautiful haunted houses, as well as an entire fantasy city (go to www.flickr.com/photos/7931559@N08/).

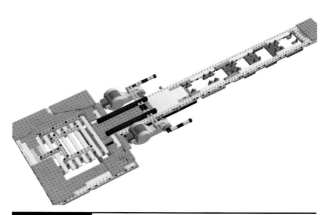

Figure 2-1 The Keytar model features Technic beams and bricks and System plates.

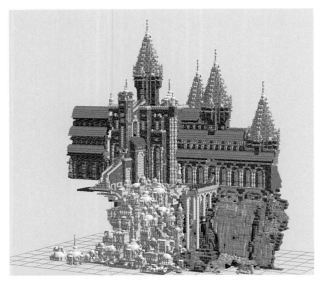

Figure 2-2   This fragment consists of 20,000 bricks, or one-tenth of the final model. (*Source:* Mike Doyle.)

A Gentleman's Tank

This outlandish tank (Figure 2-3) is part of a wide assortment of "Dieselpunk" military models built by Matthew Sylvano. This one stands out in part because of its gigantic wheels, big enough to allow the tank to traverse tall obstacles.

Figure 2-3   This fanciful tricycle tank looks especially cool with the help of ray-tracing software. (*Source:* Matthew Sylvano.)

You might also notice that this image looks good, too good to be an output from a virtual building program. Matthew used ray-tracing software that redraws the LDD files to look different (go to www.flickr.com/photos/71461420@N08/). I'll talk more about ray tracing in Chapter 10.

Death Angel Mecha

Flickr user Garry_rocks builds an endless variety of giant robots ranging from hulking humanoid brutes to alien spider robots. He even comes up with a back story for each model, fitting it into his future storyline of robot warriors. In this case, the Death Angel "is capable of ruining cities and defeating armies. It changes landscapes with its footsteps and crushes the sky with the roar of its engines. Can't you hear it? The storm is coming!"

The Death Angel is another example of a design that has been rendered after creation (Figure 2-4). This one took Garry over 22 hours to render. The quality isn't as high as he'd like,

Figure 2-4   The Death Angel Mecha does battle in the future. (*Source:* Garry_rocks.)

but his first two attempts, using a higher-quality setting, both failed (go to www.flickr.com/photos/garry_rocks/).

Chocolate Milk Maker

Another project from *Make: LEGO and Arduino Projects* is our Chocolate Milk Maker, pictured in Figure 2-5. Like the Keytar mentioned earlier, the design mates LEGO parts with non-LEGO parts. This presents a problem in depicting the contraption in LDD—for instance, the squeeze bottle of syrup, held in the clawlike gripper at the top left of the figure, has no analog in the bricks palette of LDD.

I think this is version 3 of the robot, but I built each iteration in LDD so that we could always go back and study the previous versions without having to keep them physically built.

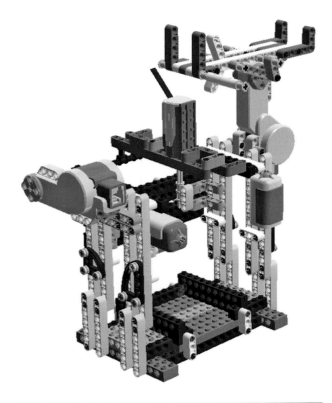

Figure 2-5　The Chocolate Milk Maker uses a squeeze bottle that can't be depicted in LDD.

A15 Comet

Paradox Kid builds lots of military models in LDD, including the A15 Comet anti-aircraft system seen in the lower right-hand corner of Figure 2-6, with the other model being a Comet reconfigured as a trench digger.

The builder likes to remix one model in many different configurations. For instance, he also created a light-tank version of the A15. His Flickr feed is chock full of endless variants of Leopard tanks and Iroquois choppers (go to www.flickr.com/photos/47503615@N04/).

Figure 2-6　The A15 Comet comes in different configurations. (*Source:* Paradox Kid.)

Alien vs. Predator Diorama

Mecha designer Garry_rocks also built this cool diorama (Figure 2-7), demonstrating that a LDD model need not be a discrete object like a car or robot. Even landscapes can be built, and it's amazing how much story can be told.

In Garry's scene (pulled from the *Alien vs. Predator* movie universe), the title monsters do battle in a ruined building surrounded by smashed tile and burst pipes. A design like this challenges builders because LDD defaults to neat and orderly, and you have to work a little harder to get messy and disorderly!

Figure 2-7 This diorama includes burst pipes and battling aliens. (*Source:* Garry_rocks.)

Pentagon

Pete Strege builds big, intricate LEGO models of buildings, and he takes pains to ensure that the proportions are correct. Most of his building models follow more rectilinear floor plans, but the Pentagon presented an obvious challenge—how to design outside the normal square LEGO pattern.

Pete's solution was to make the Pentagon a series of modules that sit next to each other with minimal gaps (Figure 2-8). There is a basement

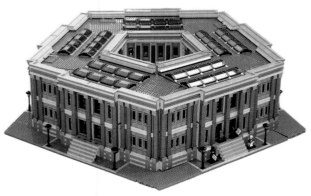

Figure 2-8 The Pentagon uses trickery to bypass LEGO's usual rectilinear format. (*Source:* Pete Strege.)

and two full floors in the model, making for 15 rooms and 45 hallways, many of which Pete has outfitted with furniture and inhabitants (go to www.flickr.com/photos/redcokid/).

Ford Pinto

Tom Netherton's Ford Pinto (Figure 2-9) demonstrates that you don't need to make giant projects in LDD—even a tiny one offers challenges. In many respects, building tiny presents an even-greater challenge because every single brick must be precisely selected and placed.

Tom's Flickr feed includes dozens of vehicle models, with everything from delivery trucks to race cars, campers, and sedans. They're all re-creations of real-world vehicles, and they all faithfully replicate the proportions of the originals (go to www.flickr.com/photos/47249637@N05).

Figure 2-9 This Ford Pinto shows that cool LDD models need not be big. (*Source:* Tom Netherton.)

Robot Chassis

This robot design consists of a motorized tank bottom packing three Mindstorms servos, with a turret-mounted, three-motor gripper controlled with Bluetooth-enabled, arm-mounted microcontrollers equipped with Wii nunchucks (Figure 2-10).

I found out while working on this project that just because you can build a model in the real world doesn't mean that you can build it virtually. Tank treads are notoriously finicky and often won't fit on a model in LDD that would work in the real world.

Pumpkin Factory

Unlike a lot of LDD models found online—particularly the really huge ones—Pete Strege's Pumpkin Factory was actually built (Figure 2-11). It comes apart in modules so that you can see the interior decoration, and there is a working Power Functions elevator.

Pete's Flickr feed shows two variants of the Pumpkin Factory, including a squat, squarish version that wasn't built, highlighting LDD's value as a prototyping medium, even for projects you intend to build.

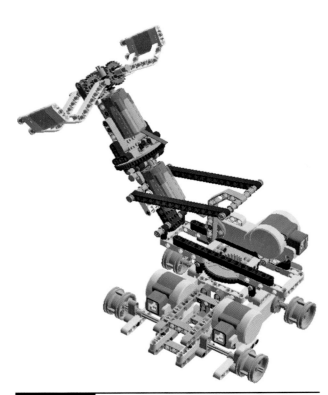

Figure 2-10 This project proves that just because a physical model can be built doesn't mean that LDD can re-create it.

Figure 2-11 The Pumpkin Factory features a working elevator. (*Source:* Pete Strege.)

Antarxa

Peer Krueger's Technic robot Antarxa is actually a bicycle because it has a front wheel and a back wheel—albeit very wide ones. It steers by retracting and extending its twin linear actuators, turning the front wheel one way or another. It uses a pair of infrared sensors to accept commands from a remote (Figure 2-12).

A robot like this underscores the lack of a proper physics engine in LDD because wouldn't it be cool to be able to "test out" a motorized robot design on screen? Peer built Antarxa and even drove it through the snow to test out its all-terrain abilities (go to www.flickr.com/photos/28134808@N02/).

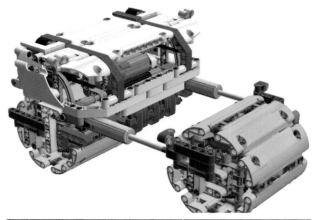

Figure 2-12 Antarxa, an all-terrain vehicle. (*Source:* Peer Krueger.)

Summary

Hopefully, the creations in this chapter inspire you to build your own creations! First, however, you'll walk through the application's capabilities and interface in Chapter 3.

CHAPTER 3

Choosing Between Software Options

LEGO Digital Designer (LDD) (Figure 3-1) will top many fans' list because it's the company's offering. It's official! The program offers a robust building system with an intuitive flair, the company supplies virtual equivalents to all the parts you want, and it has a great building instructions function to help other people re-create your projects.

As you'll learn going forward, however, sometimes fan-made alternatives have a lot to offer as well. In this chapter you'll get a clear, level-headed, and informative breakdown of the three major players and how you can differentiate them. I'll describe each platform's system requirements, ease of install, the completeness of their brick palettes, and the

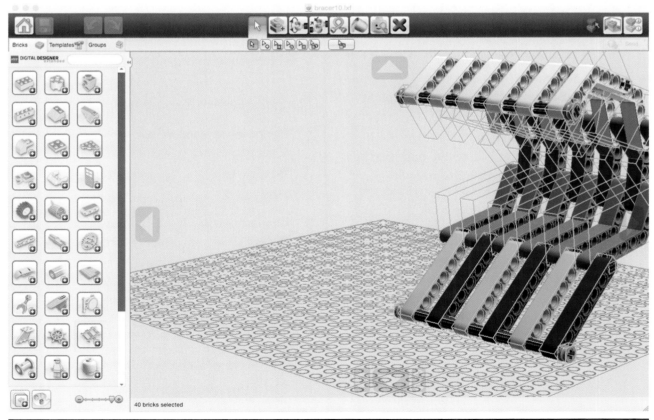

Figure 3-1 LDD in action, with the light-blue boxes connoting a multiple selection.

support and community you can expect as an adoptee.

General Editor Limitations

In general, editors (virtual LEGO building programs) share a set of attributes that keeps them tied to the physical LEGO product:

- **You can't change the size of an object, zooming notwithstanding.** All parts remain standard LEGO sizes. The virtual bricks share precisely the same dimensions as the plastic bricks found in any LEGO set.

- **You can't alter the shape of an object or change it to a different part.** Once you place a 5M beam, for instance, you can't click on it and change it to a 7M beam.

- **You can't create your own colors or use any color beyond standard LEGO element colors.** You can, however, apply colors to parts in combinations not found in existing LEGO products, such as a bright-green snowplow, which I'm pretty sure isn't found in any set.

- **The models don't respond to gravity; that is, creations that look great on screen might collapse if built in the real world.** This may not matter depending on your reasons for using the program! If you're building virtually as a prelude to translating your design to the physical world, it could be a problem.

- **LDD may not have all the latest parts.** As a non-profit-generating product, LDD won't receive updates very often; it's simple economics. LDraw and Mecabrick are dependent on fans designing the bricks in question.

System Requirements

Let's get this one out of the way. You can't build if you can't load the software! This section describes the major editors and how much computing power it takes to run the software.

LEGO Digital Designer

Typical of virtual LEGO-building software, LDD's technical requirements won't tax your computer too much. I'll include more detail on system requirements in Chapter 4, but chances are it will run on your computer if it's not too old. The big take-away involves the Linux version of LDD—or lack of it. As mentioned previously, if you want to build with virtual LEGOs and you're on a Linux machine, you won't be using LDD without special emulation software. An *emulator* allows a computer running one operating system to run a second operating system and a program intended for that operating system. I'm not an expert on emulators and would suggest that you do your own research to find the best one.

Minimum System Requirements for a PC

- **Operating system:** Windows XP, Windows Vista, Windows 7, Windows 8, or Windows 10
- **CPU:** 1-GHz processor or higher
- **Graphics card:** 128-MB graphics card (OpenGL 1.1 or higher compatible)
- **RAM:** 512 MB
- **Hard-disk space:** 1 GB

Minimum System Requirements for a Mac

- **Operating system:** OS X 10.10 or higher
- **CPU:** Intel processor

- **Graphics card:** NVIDIA GeForce 5200/ATI Radeon 7500 or better
- **RAM:** 1 GB
- **Hard-disk space:** 1 GB

In addition, older Macs (the generation called PowerPCs) may download an older version of the software, an unsupported legacy version that won't overwhelm those older machines. You can find the link to this version in the fine print at the bottom of the main download page. I'll talk more about running legacy versions of LEGO-building software in Chapter 4.

LDraw

As mentioned earlier and very likely will be mentioned again, LDraw is the library of parts that are used by separate design programs to simulate the build process. Typically, this complicated arrangement is simplified as LDraw. For the purposes of this book, I'm featuring two of these design programs, LDCad, which may be used with Windows and Linux computers, and Bricksmith, a Macintosh application.

LDCad System Requirements

The LDCad site specifies the following technical specs:

- Windows XP or higher/Linux GTK 2 platform
- AMD Thunderbird or Intel Pentium 4 or higher
- 512 MB of memory
- OpenGL 1.1 support but preferably OpenGL 2.0 or 3.0

Ideally, the computer should have the following specs:

- Windows XP or higher/Linux GTK 2 platform
- A multicore CPU
- 2+ GB of memory
- VGA card with OpenGL 3.0 support with 512+ MB of onboard memory

Bricksmith System Requirements

Go to www.ldraw.org/help/getting-started/mac.html.

Because Macintosh computers are all manufactured by one company, they have a more uniform set of stats, and therefore, it's easier to specify a computer. Simply choose any computer running OS X 10.3 (Panther) or more recent, and it will run Bricksmith.

Mecabricks

It's a website. It turns out that you can run Mecabricks if you have a HTML5-compliant browser such as Chrome, and that's pretty much it!

Ease of Installation

Another thing to consider when comparing the applications, ease of installation definitely can be a factor if you are not technically adept or simply don't want to fiddle around with a complicated installation.

LDD

LEGO makes it easy to download and install LDD. There are several steps to the process and at least one opportunity to make a mistake, as

I'll detail in Chapter 4. However, chances are that you won't have any difficulty at all. It's designed for kids to download with minimal help from adults.

LDraw

At first glance, the process for installing the LDraw library and associated programs may alarm you, if only because of the sheer number of options to choose from. It is, however, super easy to select the option you want. LDraw thoughtfully provides you with a link depending on what platform you want.

For newbies, one of these download links is sure to do the trick. The only downside is that you may not get the latest version of the LDraw library. For power users, it is recommended to first download the latest and greatest LDraw library and then select a design program and download that separately.

In addition to the editing software, you'll also have to download a *renderer*, a software engine that turns the abstract lines of the design into an apparent shape. LDraw suggests a renderer and

includes one in the downloads, and I'll talk more about that software in Chapter 10.

Then there is the fact that LDraw is a fan-built product with different parts coded by an entire group of volunteers over several years. For this reason, many times new users have a very difficult time getting the software to work. If this sounds like too much of a challenge, it probably is, but if you rely on the download links on the LDraw site (Figure 3-2), you'll probably be fine.

Mecabricks

Of course, as a web app, Mecabricks wins the prize for easiest install. But don't get too cocky, Mecabricks fans! If you're offline, you can't build in Mecabricks, no matter how awesome the computer you're using.

Brick Palettes

Of consideration is whether the application has the bricks you need and whether it's organized in the way that you're expecting. I'll provide greater detail in Chapter 5, but here is an overview.

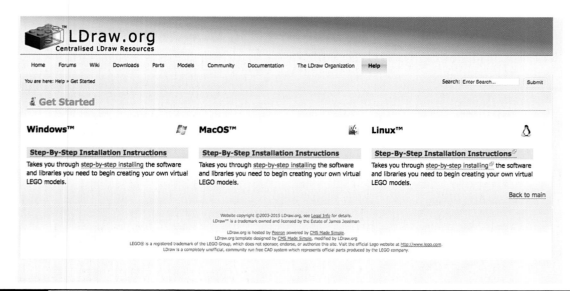

Figure 3-2   Choose among these options to install LDraw.

LDD

As the official product of the LEGO Group, one would expect LDD to have the widest assortment of bricks—and it certainly doesn't disappoint. Countless bricks (Figure 3-3) dating back to the earliest designs may be used in projects. However, the brick palettes may be difficult to navigate because LDD's interface makes it hard to distinguish between a number of similar-looking bricks.

Organizationally, the LDD brick palettes are a little confusing. They lump obscure parts in with others that don't seem to relate, making those elements remarkably difficult to find. I devote Chapter 5 to helping you find the parts you need.

Another disadvantage of LDD over LDraw is that you cannot create custom parts. That is

the price one pays for an application coded by a group of professional programmers but given away for free!

LDraw

The aforementioned LDraw library contains a breathtaking assortment of parts, broken down into a larger number of categories than either LDD or Mecabricks. For example, there is a "Propellers" category in LDraw (Figure 3-4), but propellers tend to be bunched with all transportation parts in LDD.

Not only does LDraw boast all official LEGO parts—including some obscure ones that likely will never make it to LDD—it also includes a collection of custom parts, fan-designed bricks that the company might plausibly have released but did not. In addition to mechanical

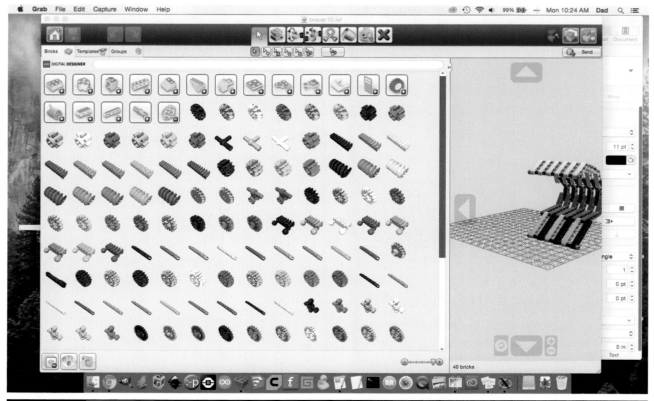

Figure 3-3 The LDD brick palettes contain countless variations of shape and color.

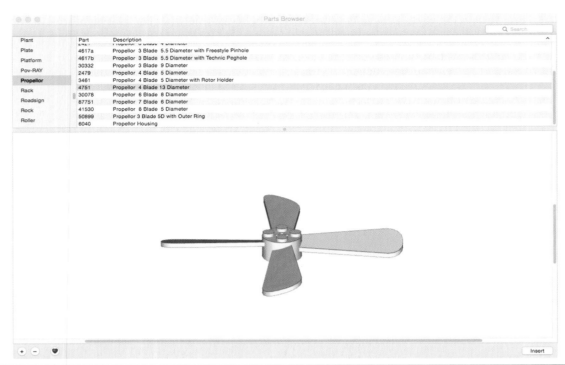

Figure 3-4   LDraw's palettes are broken down into a large number of categories with fewer parts in each.

oddities such as special bricks with an unusual configuration of mounting holes, LDraw's custom parts also include real LEGO bricks that the company never got around to digitizing. These parts aren't found in LDD but are found in LDraw. Chapter 10 offers a brief tutorial on how to build your own bricks in LDraw.

Mecabricks

Mecabricks parts (Figure 3-5) are already organized into sensible palettes, with a large percentage of bricks readily available—but not as many as LDD and certainly not as many as LDraw. However, the software does not currently support third-party bricks, and you are pretty much stuck with what the creator releases.

Another factor that might disappoint some users is that palettes cannot be reorganized or changed—and you're stuck with the same configuration as everyone else. However, it's a sensible arrangement, and I've been able to readily find every part. Finally, while Mecabricks

works fairly well with LDD, it is not compatible with LDraw, thanks to the unusual measuring system used by LDraw to draw its parts. I'll cover this in detail in Chapter 7.

Maximum Build Size

One area that may interest power users involves build size, as in how big of a model can you make with these software options? The answer is an interesting story that explains much about how the software works.

The way that virtual LEGO editors store their parts involves representing them as a series of shapes drawn according to a set of instructions. However, until they are *rendered*, or turned into something that you can see, they're just imaginary lines. The real memory draw for the computer is not the lines but the rendering. More specifically, every surface or facet of a part must be rendered separately, and this can take time and considerable computing power if you

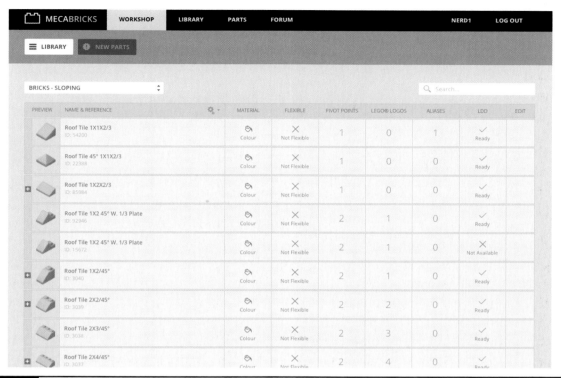

Figure 3-5   Mecabricks organizes its parts into sensible tables.

have a large number of them. As you load the parts, the software typically slows down because rendering all those parts begins to use up all the computer's memory.

So the quick answer is that typically there is no practical limit to the number of parts you can use while building. A more complete answer would be that it depends on how modern a computer you own.

Community

No one is an island, and you'll find yourself reaching out to fellow LEGO fans when you start building virtual models. Whether it's building tips or suggestions on ways to use that awesome obscure part, your online compatriots will be of great help in finding your way. Similarly, you can rely on the help of the entity releasing the software—to a degree. FAQs, help files, and discussion forums all provide

the official answers to questions. In this section I'll share some of the resources each platform has available to it. I'll expand on these items in Chapter 10.

LDD

The LEGO Group typically does not offer community architecture such as forums and chat rooms for its customers, preferring to let the product speak for itself. However, they offer one convenient feature, called the *Factory*. LEGO allows users to upload their LDD models to LEGO.com, and the models may be visible to other fans. Anyone who views the model may download the LDD file so that they can learn how to make it themselves. In prior years, LEGO had a service where you could upload your parts list, and the company would pack up all the parts and mail them to you, but the service has been discontinued.

Not all LDD resources represent corporate efforts. Some sites are fan driven, such as Eurobricks, a website frequented by some of the best builders in the world, who share tips and techniques with visitors. Another site that has caught on as a gallery of LEGO models, Flickr (Figure 3-6) allows users to share hundreds of model photographs and LDD renders, making it popular among LEGO fans.

LDraw

As an all-volunteer organization, you won't find a huge LDraw staff to help newcomers. However, the platform has a very wide community, and there are fans all around the world. Furthermore, by Internet standards, it's an *old* community. The LDraw standards were developed in the 1990s, and a lot of the original people still participate in vetting new parts and continuing to develop new software tools.

LDraw fans design new programs to work with the library, as well as new official and unofficial parts. You can use the LDraw library as a catalyst for your own creativity, such as by designing or digitizing new bricks. In Chapter 10, I'll show you how to design your own element.

Mecabricks

While new, Mecabricks already has a passionate user base as well as a very active creator. New questions posed in the forums (Figure 3-7) get answered very quickly either by the moderator or by the creator himself. In addition, many of the fans are former LDD and LDraw users and can answer generic questions about building techniques. Don't be surprised to see changes happening with this platform.

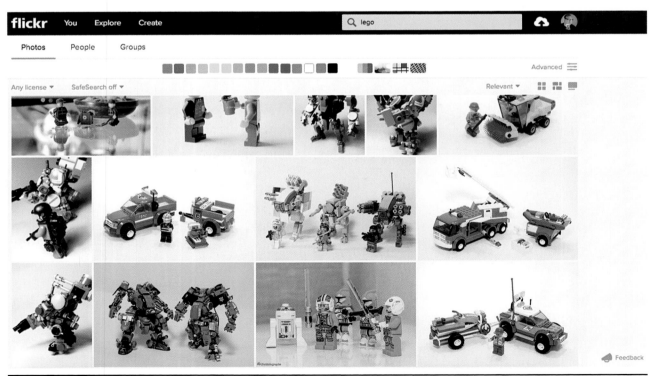

Figure 3-6   A photography site, Flickr has become a home for LEGO builders.

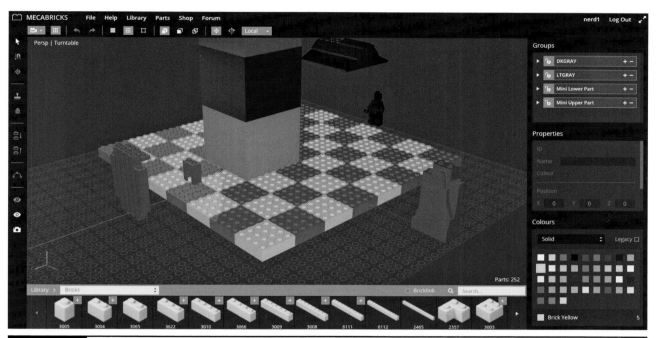

Figure 3-7 Young and hungry Mecabricks users help you to build online.

Summary

With the differences and similarities laid bare, it's time to choose your software package. Chapter 4 will guide through each step of the download and installation processes for those packages.

Installation and Overview of LDD and LDraw

THE MOST LOGICAL FIRST STEP to using a virtual LEGO building program probably involves downloading and installing the software (Figure 4-1), so let's begin there! I'm excluding Mecabricks from this chapter because it's a website, with all the advantages and disadvantages one would expect. There are times when using a web app makes things easier, and not having to install software is definitely one of them.

Downloading LEGO Digital Designer

Very likely you've downloaded programs and applications before, so it should be a cinch to get LDD installed.

Download the LDD Application

The software may be downloaded only from LDD's website. Simply navigate to

Figure 4-1   The LEGO Digital Designer (LDD) download page guides you through the process.

http://ldd.lego.com/en-us/download/ and follow the instructions.

Choose Your Platform

If you have a Mac or PC, now is the time to select the platform you'd like to download.

Figure 4-2 shows the page and also repeats the system requirements. As I mentioned in Chapter 1, other operating systems besides PC and Mac *are not supported*. This includes the commonplace open-source Linux OS, which is a pity because a lot of LEGO nerds I know like Linux as well.

System Requirements

Before you go too far, however, let's take a quick look at LDD's minimum system requirements to successfully use the program on your computer. Note that these may have changed if the software has been updated since the publication of this book. You can see the requirements online on the LDD's download page.

PC

Operating system: Windows XP or better

CPU: 1-GHz processor or higher

Graphics card: 128-MB graphics card (OpenGL 1.1 or higher compatible)

RAM: 512 MB

Hard-disk space: 1 GB

Mac

Operating system: OS X 10.6.8 or higher

CPU: Intel processor

Graphics card: NVIDIA GeForce 5200/ATI Radeon 7500 or better

RAM: 1 GB

Hard-disk space: 1 GB

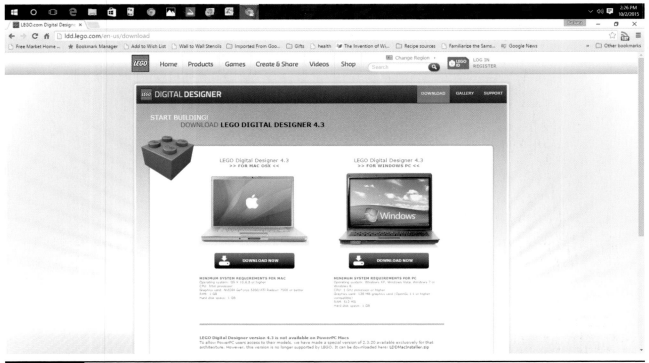

Figure 4-2   Choose a platform.

Downloading LDD for PC

The instructions vary based on system type, so this section guides you through Windows PC download and setup.

1. Choose the save location for the setup wizard's executable, as seen in Figure 4-3.

This isn't the final destination of the LDD program, and folks usually just drop it onto their desktop.

2. Launch the setup wizard from the desktop. Figure 4-4 shows the initial screen.

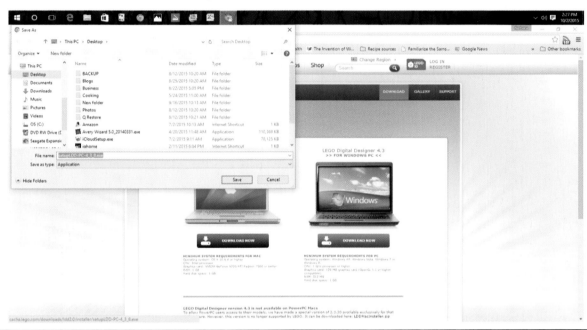

Figure 4-3   Choose a save location for the installer executable.

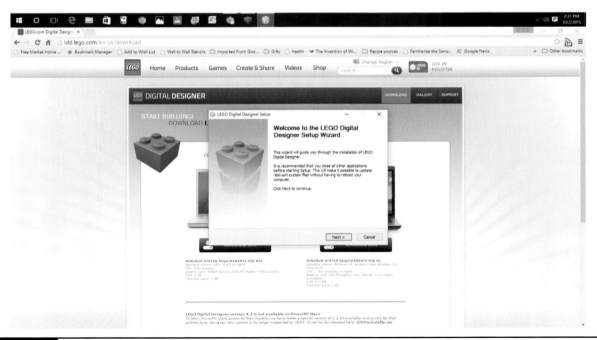

Figure 4-4   The setup wizard will take care of installing LDD.

3. License agreement. Most likely you won't read the fine print in the license (seen in Figure 4-5). Click away!

4. Choose components, as shown in Figure 4-6. Most of the time the default of "everything" is fine. Just click through to the next item. This step, in my opinion, is the only one that has any chance of confusing you.

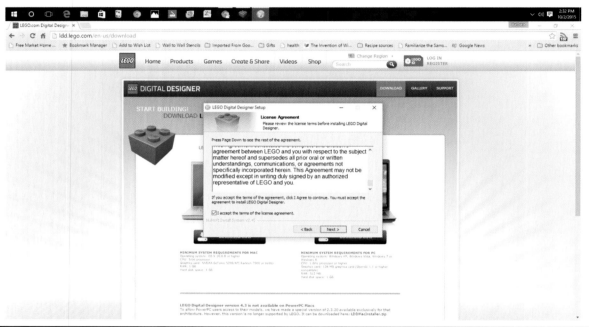

Figure 4-5   Click through the license agreement.

Figure 4-6   Choose components.

5. Before installing, the software checks the system against its list of requirements. Figure 4-7 shows the software working through this.

6. Choose the install location, as seen in Figure 4-8. I'd go with the default, which drops LDD into Program files.

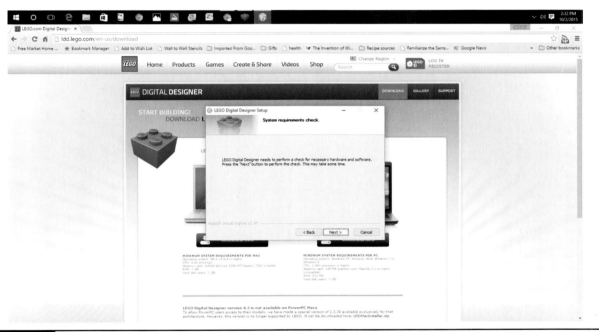

Figure 4-7   Checking the system requirements.

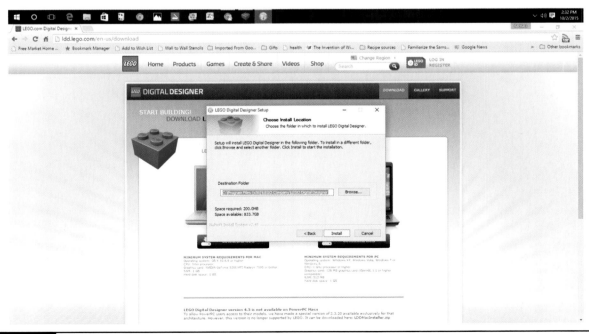

Figure 4-8   Saving LDD into Program files.

7. Begin installation! It will take a couple of minutes. Figure 4-9 shows the tantalizing green progress bar.

8. Finish and run. You're so close! Click on the "Finish" button (Figure 4-10), and start playing!

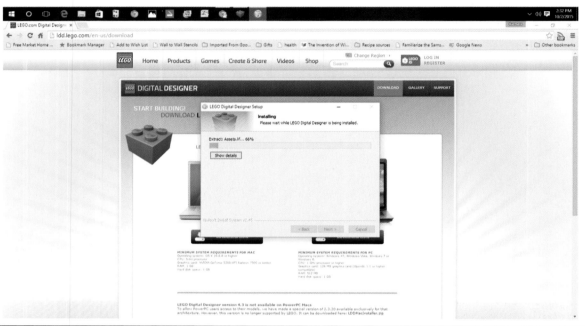

Figure 4-9   The green bar marks the installation's progress.

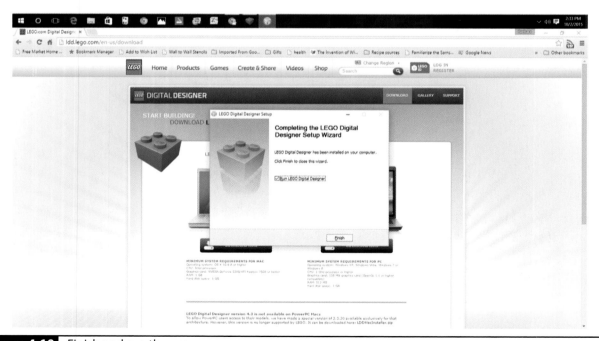

Figure 4-10   Finish and run the program.

Downloading LDD for Mac

You'll find that the procedures for downloading to a Macintosh computer closely resemble those for a PC. There are, however, a couple of tricky parts. Let's go over the steps:

1. Download and unarchive the package. The package contains the installer file. When it has been downloaded, launch the installer to see the screen in Figure 4-11.

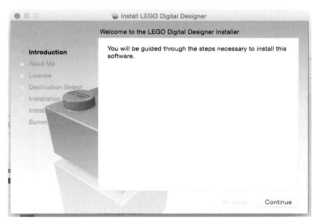

Figure 4-11   Launch the installer and follow the directions.

Note: There is a common security hurdle encountered by folks installing on a Mac. The Mac won't install until you confirm that you're okay with downloads from a non-Apple site. Simply go to the "System Preferences" menu, and then click on the "Security & Privacy" submenu. At the bottom of the screen it lists your preferences with regards to app downloading. If you select "Allow apps downloaded from: Mac App Store," you won't be able to install. The other two options permit installation, so choose one of these to proceed. If you're doing this after being told by the system you can't install, the Security & Privacy page will specifically mention that file and ask if you want to specifically allow it without changing your security settings.

2. Read the fine print. Once again, you're faced with having to read and digest several pages of legalese, as seen in Figure 4-12. Click "Continue" to proceed to the next step.

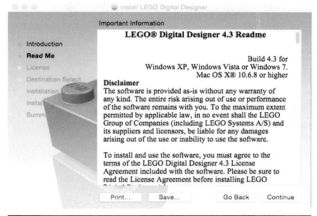

Figure 4-12   Read and sign the licensing agreement.

3. Select a download destination, as seen in Figure 4-13. Most of the time the default of "Applications" will suffice.

Figure 4-13   Select a download destination.

4. The installation proceeds, with a status bar (Figure 4-14) showing how much time is left.

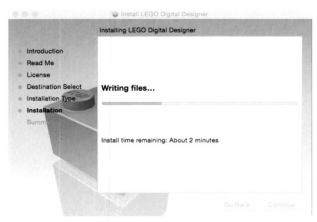

Figure 4-14 The status bar shows much time is left.

5. Launch the application (Figure 4-15) and start playing!

Downloading LDraw

As I mentioned earlier, lesson number one in working with LDraw is that the LDraw library is not the same thing as the CAD program (like Bricksmith) used to build models. This section therefore describes how to download both.

The editor developers wisely offer "quick install" options for non–power users who might not care about maxing out the experience and just want to get up and running as quickly as possible.

Choose Your Package

LDraw's "getting started" page (www.ldraw.org/help/getting-started.html) invites you to select the platform for which you're downloading. The links connect you to subpages (Windows' is seen in Figure 4-16) that guide you through selecting which editor you want to download. As you may recall, several editors are actually available; I just choose LDCad as the best for Windows and Linux and Bricksmith for Mac.

LDraw editors are useless without the parts (Figure 4-17). If you elect to download LDCad by itself, you absolutely must get the library

Figure 4-15 Launch the application.

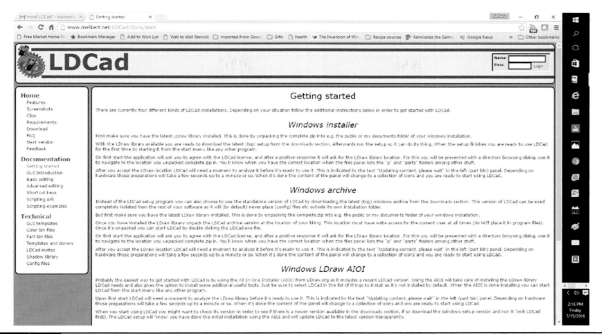

Figure 4-16   Choose which package you want to download.

first. The link may be found on LDraw's website: www.ldraw.org/parts/latest-parts.html. If you notice, there are links to get just the latest parts, the complete library, or, even more helpfully, a Windows installer for the complete library that puts everything where it's supposed to be.

Quick-Install Option

The LDraw organization rightly focuses on the library, their raison d'etre, and if you learn about the software from the LDraw.org site, you'll learn a lot about the library but little about the editing software. The reason for this is that

Figure 4-17   Get the LDraw parts separately if you're worried about having the latest.

the LDraw organization doesn't want to play favorites with the editors. They would like to work with every single editor in existence.

If you visit the editors' websites, however, you'll find a quick download link that grabs not only the editing software and the library but also third-party applications such as rendering engines, which I'll describe in Chapter 10. While convenient, the disadvantage to these quick downloads is that you may not get the latest and greatest library of elements. Almost everything you need, probably, but not *everything*. The only way to be sure of grabbing everything is to manually download the latest library from LDraw.org's "Latest Parts" page and get the editor only from the application's site.

However, I shouldn't scare you. If you click a download link for either Bricksmith or LDCad, you're almost certain to get the latest parts.

Downloading LDCad for Windows

LDCad for Windows includes an all-in-one installer (AIOI), making installation of the various components a breeze. The advantage to this is obvious: you get everything you need in one download. On the downside, you may get software that you don't need. Similarly, it's possible to miss out on a LDraw parts release if you download an older installer. However, those possibilities are pretty remote, so I suggest using the AIOI.

1. Visit the "Downloads" page (Figure 4-18). This page serves as your gateway to the world of LDCad. There are links for both Windows and Linux versions of the software, but let's focus on the former for now. Choose either to download the LDCad installer (called "Setup") or a folder with all the files in it (labeled "Archive") for you to install manually. You want "Setup" (www.melkert.net/LDCad/docs/start).

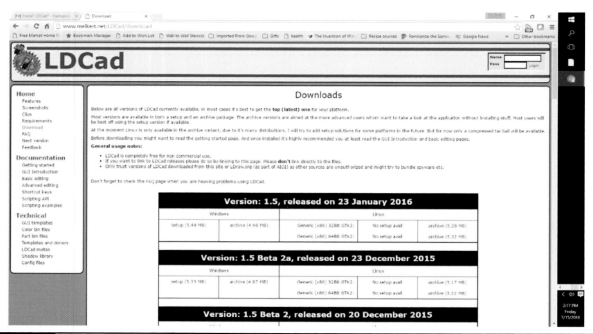

Figure 4-18 Download the latest version of LDCad from here.

2. Save the LDCad executable. This file (Figure 4-19) contains everything you need to start building and drops each program into the right folder. You can put this executable on your desktop for future deletion.

3. Select LDCad's installation folder, as seen in Figure 4-20.

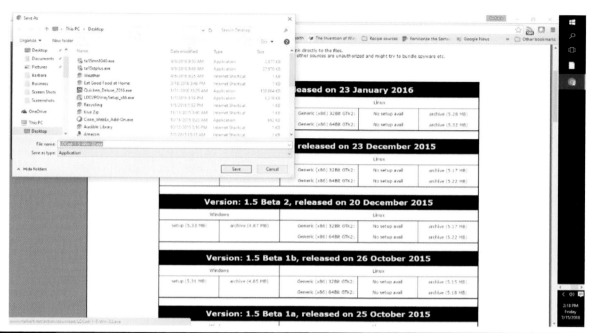

Figure 4-19   The LDCad installer contains everything you need.

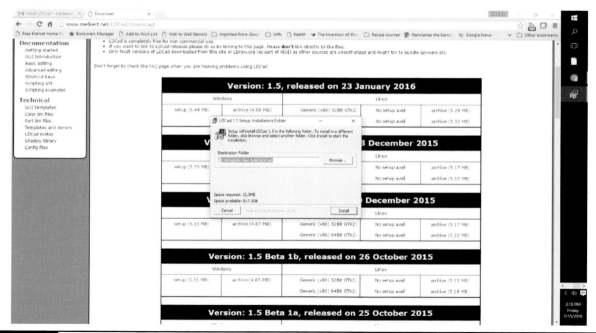

Figure 4-20   Select the "Installation" folder.

4. Save the LDCad installer, and launch the executable (Figure 4-21), which will do the actual installation of the software. Click on "Save," and you'll move onto the next step.

5. Set up the LDraw installation folder, as seen in Figure 4-22. Having clicked on "Save," you'll be prompted to choose a folder into which to save the LDraw library files.

Figure 4-21 Launch the LDCad installer.

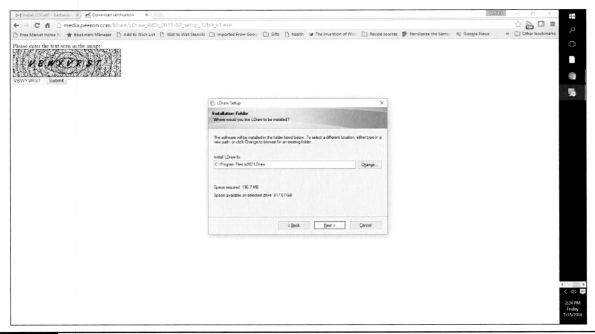

Figure 4-22 Choose the LDraw installation folder.

6. Uninstall the previous version (Figure 4-23) if you still have an older version of LDraw's all-in-one installer on your computer.

7. Approve the LDraw license agreement. The fine print (seen in Figure 4-24) explains the conditions under which the software has been released.

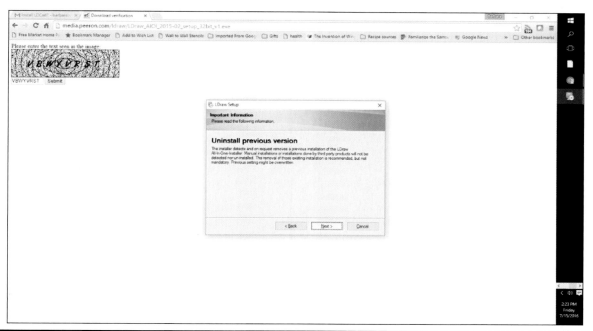

Figure 4-23 If you have the older install file, uninstall it.

Figure 4-24 Read the license agreement, and then click on the "Accept" button.

8. User information (Figure 4-25) prompts you to create a nickname for yourself that will be used to brand your LDraw projects.

9. Select "Packages," as seen in Figure 4-26. You're not just downloading the editor;

LDCad gives you the option to download third-party applications such as LPub3D, which creates building instructions for re-creating your models, and LSynth, which helps to manage bendable elements.

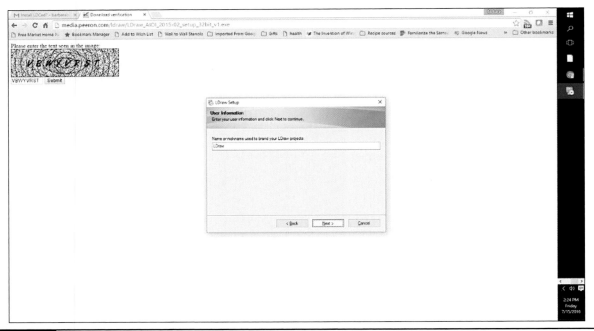

Figure 4-25 Come up with a screen name for yourself.

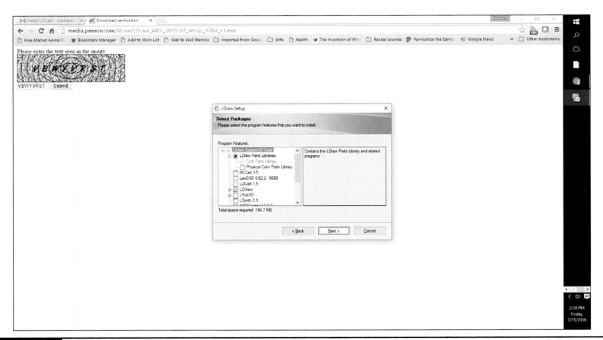

Figure 4-26 You're not just downloading the parts and editor.

10. Choose or create the LDraw parts library installation folder (Figure 4-27), and the system will unpack the library there.

11. Designate a folder for shortcuts, as seen in Figure 4-28.

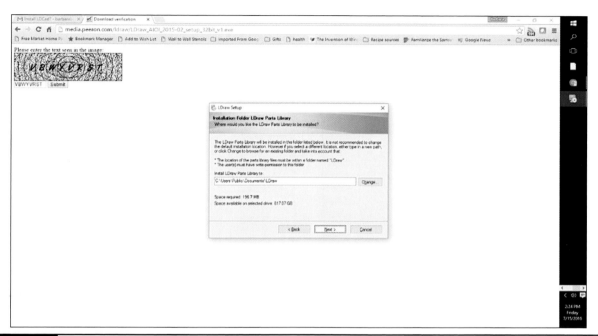

Figure 4-27 Choose where to put the LDraw parts library.

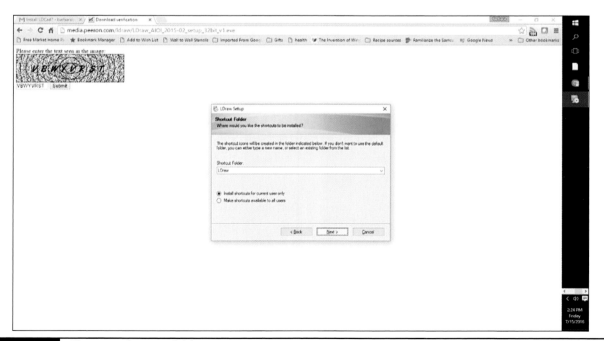

Figure 4-28 Choose a shortcut folder.

12. Select installation options (Figure 4-29), such as whether to create a desktop icon and whether to open up a model on installation.

13. The installer has enough information to complete the install, as seen in Figure 4-30.

14. Finally, the installation finishes up (Figure 4-31), and you're ready to start building.

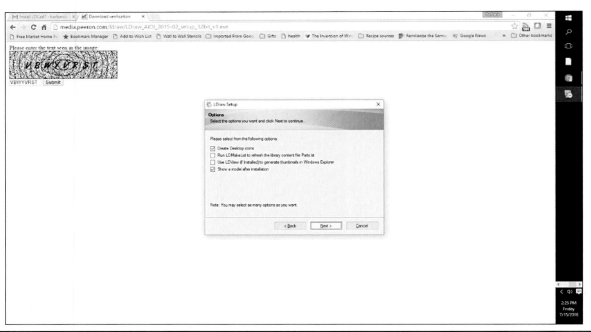

Figure 4-29   Select installation options.

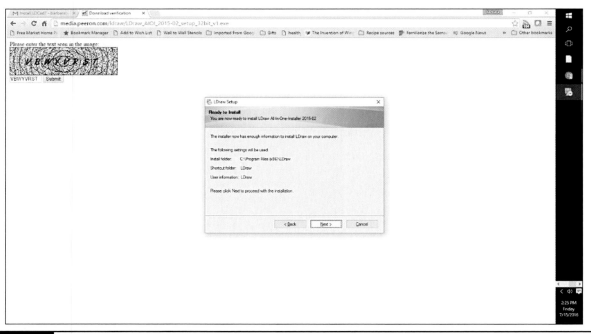

Figure 4-30   Ready for installation.

Figure 4-31 Installation is successful. It's time to start building!

Downloading LDCad for Linux

Next, you'll learn how to install LDCad on a Linux machine. This section is a little vague because Linux actually consists of many different variants, called *distributions* or *distros*, and they all work differently. An application that works on one distro won't work on another. Because of this, many Linux users won't be able to use even the Linux-friendly LDCad. Nevertheless, the following is a general guide to installing LDCad on a Linux machine:

1. Download the latest version of the LDraw parts library, and install it as you would normally—simply unpack the .zip file into a folder. The Linux version of LDCad does not come with an all-in-one installer, so you'll have to do it all manually, including downloading the LDraw library.

2. Download the latest version of the LDCad archive. You may find both 32- and 64-bit versions of the software in the "Downloads"

section, and which one you select depends on what Linux distribution you use.

3. Uncompress the tar ball into the chosen location, but make sure that the folder offers write access to all possible users.

4. Run the genDskEntry.sh script to generate an LDCad desktop file that can be put in a task bar or menu.

5. Launch the application, and okay the licensing agreement. LDCad will prompt you to enter the location of the LDraw library.

6. After the LDraw location has been selected, LDCad will spend a minute or so "updating content," and the LDraw parts won't be visible. When it's done, the part icons will be visible, and you're ready to begin.

Downloading Bricksmith for Mac

While there is no slick all-in-one installer for the Mac, chances are you'll find installing Bricksmith to a Macintosh computer blissfully easy (Figure 4-32).

1. Download the latest LDraw parts library from www.ldraw.org/parts/latest-parts.html or else be prepared to download the library with the application.

2. Download the application from http://bricksmith.sourceforge.net/, and unzip the program archive. Alternatively, you could download a package that includes both the application and the library.

3. Drag the application file into the "Applications" folder.

4. Be prepared to specify the location of the LDraw parts library.

5. You're done!

Bricksmith
virtual Lego modeling for your Macintosh
By Allen Smith

Resources

LDraw.org
Learn about the LDraw library of virtual Lego bricks, upon which Bricksmith is based.

LPub
Make printed instructions from your models.

LDView
Fly through your models and export them to POV-Ray raytracer format.

LSynth
Add fiendishly complicated flexible hoses to your creations.

POV-Ray
Turn your LDraw models

Bricksmith allows you to create virtual instructions for your Lego creations on your Mac. The magic is based on the LDraw library, a collection of 3D models of Lego building blocks created by enthusiasts from around the world. With Bricksmith, you'll never have to worry about running out of parts!

Bricksmith offers:

- drag-and-drop building
- thousands of parts at your fingertips, with no limitations on which ones you can use
- a full Lego color palette
- support for steps and submodels, to document exactly how to build your creation
- undo, cut, copy, and paste
- multiple viewports
- a Minifigure Generator
- drag-and-drop flexible element generation

Soak in the wonderment. View the screenshot.

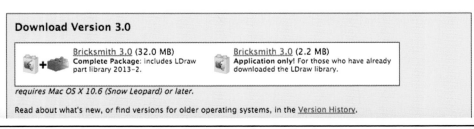

Download Version 3.0

Bricksmith 3.0 (32.0 MB)
Complete Package: includes LDraw part library 2013-2.

Bricksmith 3.0 (2.2 MB)
Application only! For those who have already downloaded the LDraw library.

requires Mac OS X 10.6 (Snow Leopard) or later.

Read about what's new, or find versions for older operating systems, in the Version History.

Figure 4-32 Bricksmith gives you the option to download either the application alone or with the LDraw library as well.

> *Note:* As with the LDD install, the Mac won't install until you confirm that you're okay with downloads from a non-Apple site. Simply go to the "System Preferences" menu and click on the "Security & Privacy" submenu. At the bottom of the screen it lists off your preferences with regard to app download. If you selected "Allow apps downloaded from: Mac App Store," you won't be able to install. The other two options permit installation, so choose one of these to proceed. If you're doing this after being told by the system you can't install, the "Security & Privacy" page will specifically mention that file and ask if you want to specifically allow that one without changing your security settings.

Legacy LDraw Software

Getting the most recent edition of a piece of software remains the obvious default, but many users actually want an older version. Maybe they just liked it better, or perhaps they have an old computer that doesn't run the new software. If this is the case with you, definitely check out the various legacy downloads.

LDraw Parts Library

The LDraw Project has been active since the 1990s, and this means that there are countless earlier versions of the designs. Their "Part Updates" page records releases as far back as October 1997. You should note, however, that you're unlikely to get anything more computer friendly. LDraw part files (tagged .dat) are extremely small and stored in the form of text files, which don't take up much space—the entire 7,000+ brick library barely takes up 5 MB.

What you'd get with that older parts library are a smaller assortment and more deprecated bricks that have been updated with newer designs.

Bricksmith

Bricksmith version 1.2 was released in 2007 and works on MacOS X 10.3.9, making it usable on that legacy computer you have in your basement. Of course, as an older edition, you'll find missing features and unresolved bugs because, well, it's an older version. You can find earlier versions of the software dating back to 2005, but the latest is from 2013. See the "Version History" page to see all your options (go to http://bricksmith.sourceforge.net/VersionHistory.html).

LDCad

LDCad is a newer app than many others, but there are a number of older versions available for download dating back to Version 1.0 beta 1 in 2011.

MLCad

One of the most popular LDraw parts editors of yesteryear is MLCad, created by Michael Lachmann and formerly the software that everyone used in the LDraw scene. The most recent version dates from 2011, and it is no longer developed or supported. A few diehard MLCad fans still use the software, however (go to http://mlcad.lm-software.com/).

LDD

LDD gets in on the legacy software angle. Not only is LDD version 4.3 (the latest) usable on even a Windows XP machine, but the company has an older Macintosh version (version 2.3.20) intended for the old non-Intel versions of the hardware (go to http://ldd.lego.com/en-us/download/).

Summary

This was a useful chapter because there are so many different ways to build with virtual LEGOs, and you should be ready to go with the software. In Chapter 5, you'll tackle perhaps the most important part of the program: the libraries that contain all the LEGO shapes you need to build.

Brick Palettes

THE INTERFACE ELEMENTS that store all the brick shapes are called the *brick palettes* (Figure 5-1). They can be pretty intimidating because LEGOs have a bazillion parts, comprised not only of ordinary bricks you'd use every day but also numerous obscure ones that can sometimes be confused for the commonplace ones. Adding to the mess is the fact that some palettes often contain parts that ought to have been placed in another palette.

In this chapter I'll guide you through the various palettes and make some observations about each one. First, however, I want to talk about two very interesting ways to customize the palette experience to make it work better for you: templates and groups.

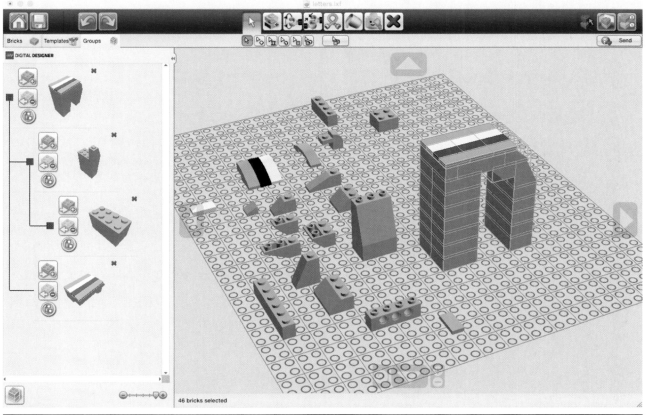

Figure 5-1   These folders store various subsets of LEGO elements.

LEGO Digital Designer (LDD)

Let's get on with the meat of the chapter: the gloriously jam-packed brick palettes, one of which can be seen in Figure 5-2. Each palette is displayed as one of a series of folders than can be expanded to show thumbnails of the various parts. From there you simply have to find the element you want (sometimes a tall order!) and click on it to create an instance on the workspace, already selected.

Identifying an Element

While the sheer number of parts may overwhelm a casual viewer, there are some basic tactics you can use to determine which part is which.

- **Brick name.** Each element has an "official" name. I put official in quotes because you may find a given element is called one thing in one LDD and another thing in other LEGO publications. Still, it often helps to differentiate two different nearly identical

parts. Simply hover the mouse pointer over a thumbnail to see a yellow pop-up showing the element's name.

- **Element number.** Each LEGO brick has a part number stamped on it, and virtual bricks include their part numbers in their description fields. When in doubt, Google the part number and confirm.

- **Rotate the element.** Hold down the CONTROL key on your keyboard, click-and-hold on a brick, and then move the mouse around. This will rotate the thumbnail around, letting you see all sides of the part. Sometimes seeing an element from different angles casts light on whether it's the one you want.

- **Expand thumbnail size.** You can increase the size of the thumbnails by sliding the icon size slider and expanding the windowpane to fit them all in the view. This will give you more detail to use in making your decision.

- **Switch themes.** As mentioned earlier, the palettes differ from theme to theme. For

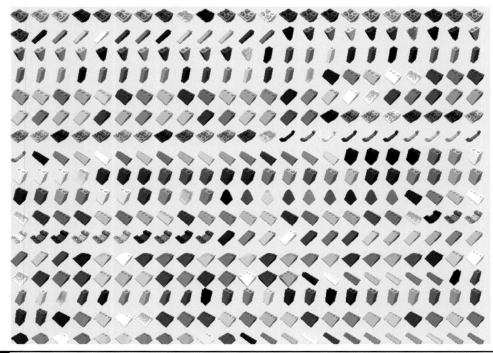

Figure 5-2   Don't get overwhelmed by all the parts available to you!

instance, the Mindstorms theme mostly just has Mindstorms parts. Sometimes the reduced diversity can offer some clarity in parts selection. For instance, in Extended, there are a million nearly identical wheel rims in one of the palettes. However, if you look at the theme in Mindstorms, there are only four rims from which to choose, two of which are different-colored versions of the same one. This makes it easy to find that perfect Mindstorms wheel.

Exploring the Palettes

There are thousands of parts in LDD, so ultimately you'll have to check them out yourself. The following is a tour of the palettes and some of the parts that may be found in them. In my descriptions I'll sometimes mention one or two parts of note per palette, focusing on stuff you'll use all the time; "head scratchers," where a part was evidently put in the wrong palette; or just a fun part that I like.

> TIP: In my coverage of the brick palettes, I'm mostly just showing the Extended palettes. Here's why: a lot of the basic and Mindstorms palettes are repetitive—look back to Figure 5-2 to see what I mean. If I showed the entirety of each palette, it would nearly be the book right there. Instead, I'm just featuring the Extended palettes because these show one of everything. In cases of printed bricks and multicolor bricks, which are depicted accurately in LDD and Mindstorms, I'll mention those alongside their Extended equivalents.

Let's dive in!

Classic Bricks

The first palette, appropriately enough, has such elements as the original 2 × 4 brick. You can see the various options in Figure 5-3.

Figure 5-3 The first palette holds variants of the classic rectangular LEGO brick.

Round Bricks

Round bricks remind you of the classic rectangular ones except, well, the obvious shape difference. As you can see from Figure 5-4, there are some seemingly random additions. None of them are particularly interesting, and they seem to have been placed there simply because one part of the element was rounded. You'll encounter this a lot when digging through palettes. Some oddities can be justified, but some are so arbitrary that it's hard to figure out why they are included.

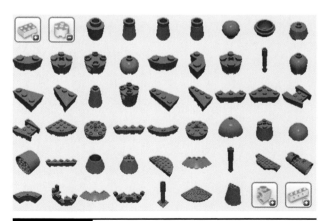

Figure 5-4 Find your round bricks in this palette.

Miscellaneous Bricks

This palette is vaguely titled, but bear with me. This palette (Figure 5-5) has a bunch of square-shaped (or approximately so) bricks that don't fit into any other category. You can see that most of them are simply a variety of the classic rectangular LEGO brick with various add-ons like balls, sockets, forks, and so on.

Figure 5-6 is an example of a part that fits into this palette's "mostly just square bricks" theme. Actually, two parts are shown, "Mountain Top" and "Mountain Bottom." These combine to create the mountain shape you see in the figure. I'm not sure why LDD can't just have a "Nature" palette with trees and rocks and flowers in it, but that's not how it worked out.

Technic Bricks

I love Technic bricks because they combine both the side holes of Technic beams along with the studs of classic LEGO System bricks. They're a great way to combine both sets into a single model. There are a couple of weird parts, but

mostly the bricks in this palette are nothing interesting. This palette (Figure 5-7) is just a repository of typical Technic bricks.

Slope Bricks

The next palette shows various slope bricks such as roofs, as seen in Figure 5-8. Like the first and

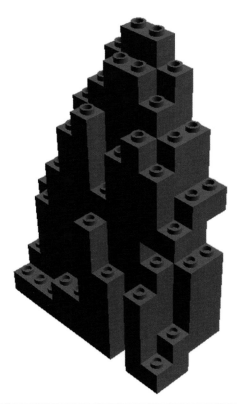

Figure 5-6   This mountain element is actually two parts joined together.

Figure 5-7   Technic bricks allow you to combine Technic beams and LEGO System bricks.

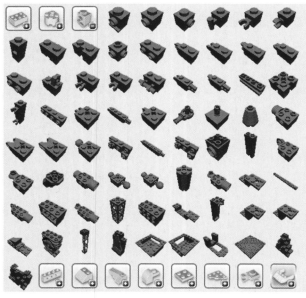

Figure 5-5   This palette seems to be a junk bin of parts.

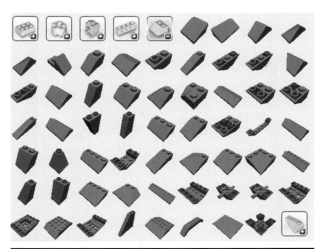

Figure 5-8 Slope bricks let you make roofs, among other things.

Control Panel Bricks

One subset of slope bricks consists of those which have been screen printed with graphics, typically control panels for spaceships and other vehicles. As mentioned earlier, you won't find graphic bricks in the Extended palettes, but you will find them in the regular LDD slope-brick palette. Figure 5-10 shows some of these bricks mixed in with the standard slope bricks.

Additional Slope Bricks

Figure 5-11 shows the next palette, which serves as a collection of overflow slope bricks.

second palettes, a lot of these bricks are as old as LEGO—the red 2 × 2 and 3 × 4 slopes resonate particularly with me because those classic parts were in fact molded in red.

As with some of the other palettes, LEGO lumped stuff together simply because parts of the element resembled parts of other bricks. In the case of the "Edged Brick w/Snap 6 × 6 × 2" pictured in Figure 5-9, the angled surfaces of this decidedly non-roof-like part was enough to get it put in this palette.

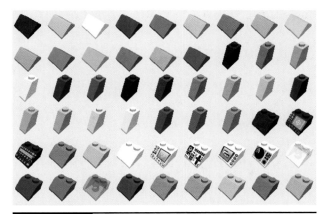

Figure 5-10 Control panel bricks add science fiction detail to any build.

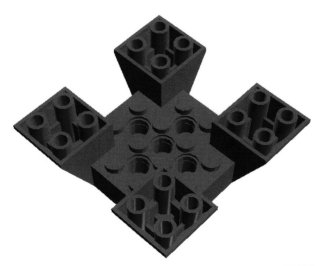

Figure 5-9 The "Edged Brick w/Snap" is an oddity in an otherwise predictable palette.

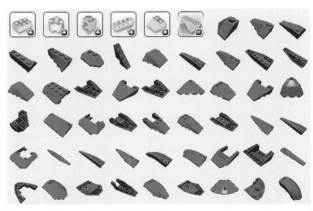

Figure 5-11 This palette includes numerous slope bricks that didn't make it into the main slope folder.

Arch Bricks

The Arch Bricks palette holds the expected arches, as well as parts with archlike parts, such as the box shapes toward the end of the parts shown in Figure 5-12, as well as the very last part, an entire building archway.

Rectangular Plates

This unexciting palette (Figure 5-13) holds rectangular LEGO plates of many different sizes ranging from 1 × 1 to 16 × 16.

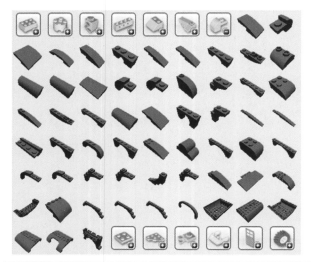

Figure 5-12 The Arch Bricks palette mostly contains arches.

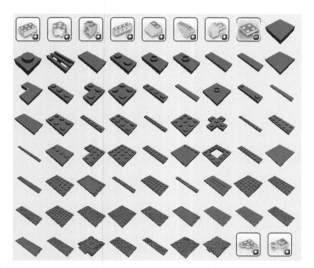

Figure 5-13 Need a flat LEGO plate? Check out this palette.

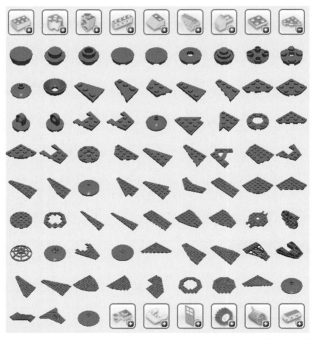

Figure 5-14 This palette holds nonrectangular LEGO plates.

Circular and Triangular Plates

Continuing on with the plates theme, this next palette (Figure 5-14) holds nonrectangular plates, including a lot of asymmetrical wing-shaped plates as well as circular ones.

Miscellaneous Plates

This is another of those palettes that serves as a catch-all for parts that can't easily be categorized. Looking at Figure 5-15, you can see the sheer variety of plates, many of which are extremely obscure.

The Miscellaneous Plates palette holds a lot of obscure parts! As is typical, there are also miscategorized parts. One example might be the "Technic Rotor w/3 Blades" pictured in Figure 5-16. It's a Technic-compatible plate with a cross-axle hole in the center and three blades with LEGO studs on them.

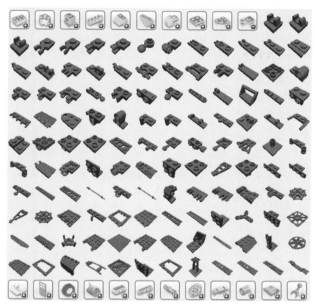

Figure 5-15 The Miscellaneous Plates palette has a lot of obscure parts.

Figure 5-16 The "Technic Rotor w/3 Blades" probably should have been put with the Technic parts.

Windscreens

The Windscreens palette holds a bunch of domes, windshields, and cockpits. It's a pretty cut-and-dried palette without a lot of odd parts. Figure 5-17 shows the variety of these parts—all opaque red, of course! Never fear, in Chapter 6 I'll show you how to make these parts transparent. In the meantime, you can view the palette in the blue-tab LDD theme to see auto-generated transparent windscreens.

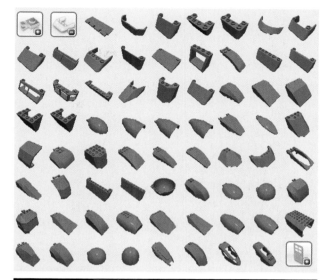

Figure 5-17 This palette holds a wide variety of vehicular windows.

Architectural Elements

This palette consists mostly of doors and windows, as seen in Figure 5-18. There are other elements as well, including gates, cupolas, casements, and wall sections. This is a fairly clean palette without a lot of elements added for no apparent reason.

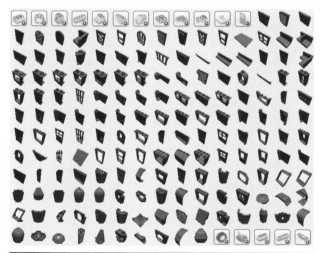

Figure 5-18 Need to dress up your house? These windows and doors will fit the bill.

Figure 5-19   This steeple consists of two halves that attach together.

One of my favorite elements is a two-part Victorian steeple (Figure 5-19), perfect for building LEGO haunted houses and other ornate structures. Unlike a lot of roof bricks in LEGOs, it's made of two large parts, giving you a different aesthetic than one you might make with a couple of dozen slope bricks.

Wheels

I find this palette the most frustrating because a lot of the parts look alike. I already mentioned the example of the wheel rims. They all look like knobby cylinders! Figure 5-20 shows what I have to deal with. Imagine trying to find a specific tire and a specific rim that fit together. It can be a headache!

One of the fun contents of this palette is a variety of chain links. These are individual plastic parts that link together with a bunch of others just like it. One of these assemblies is shown in Figure 5-21. Simply create a bunch of instances and link them all together.

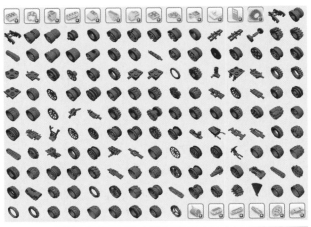

Figure 5-20   The Wheels palette has every wheel you could hope for and a few extras besides.

Figure 5-21   This chain was made up of several smaller parts.

Mindstorms and Technic

This palette (Figure 5-22) holds Mindstorms and Technic motors, sensors, and controller bricks, as well as the wires to connect them. Note that if you want to find these multicolored parts in the correct LEGO color combinations, simply look for them in the right palette in the Mindstorms theme.

Technic Beams

If we are looking for Technic parts, this palette is the key. It holds straight and angled beams (Figure 5-23) compatible with the Technic and Mindstorms lines.

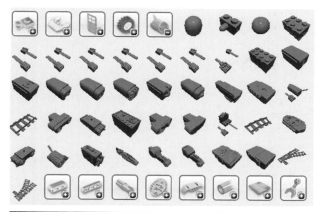

Figure 5-22 Find your motors and sensors here.

Figure 5-23 This palette holds Technic beams.

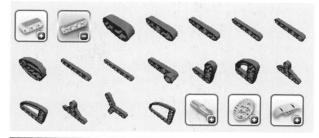

Figure 5-24 Half-width Technic beams are often used to reinforce robots.

Thin Technic Beams

This palette (Figure 5-24) holds yet more Technic beams, but these beams are half the standard thickness. These bricks are often used for reinforcements for Technic robots.

Pegs and Axles

This is probably my number 1 palette when building using Technic elements. I call it "Pegs and Axles" chiefly because those are the first two types of elements. However, if you look at Figure 5-25 you can see some decidedly nonpeg and nonaxle shapes!

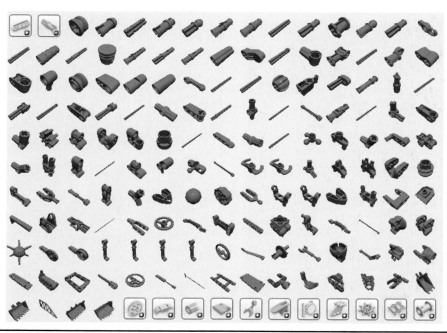

Figure 5-25 Pegs, axles, and a whole lot more besides.

This palette truly has the most oddities and misplaced parts, which is all the more peculiar when you realize that you'll need parts from this palette for nearly every Technic and Mindstorms robot. Maybe it's easier to say that it serves as an "all of the above" palette for those two product lines.

Some of the most common bricks I use are the beam frames pictured in Figure 5-26. They're buried in the Pegs and Axles palette, but if you build with Technic parts, you'll use them all the time.

Other go-to parts are the Technic beams with built-in pegs (Figure 5-27), making them great for reinforcing Technic models.

Figure 5-27 Two variants of a Technic beam with built-in pegs.

Gears

Continuing on with the Technic theme, this palette (Figure 5-28) holds all the various gears found in LEGO products, as well as associated parts such as a gearbox and gear-operated turntable.

Figure 5-26 Beam frames are great for reinforcing a robot chassis.

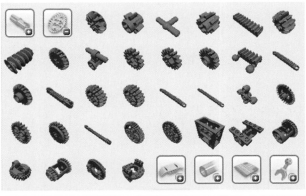

Figure 5-28 LEGO's gear offerings may be found in this palette.

Car Parts

This palette gathers together the various fenders, spoilers, and other parts of a car's body. The only real oddity is the "Bionicle Eye" (the first part in Figure 5-29) that finds its way into a lot of sets but isn't really car related.

Flex Elements

Here's a fun palette. This one mostly covers flexible parts, including rubber axles, pneumatic tubes, and cosmetic parts such as strings and nets—you can see these parts in Figure 5-30. This palette also includes some seemingly bizarre bricks, but these are parts of the pneumatic set that go with the tubes. I'll explain how to manipulate flexible elements in Chapter 7.

Street Plates

This specialized palette simply holds plates designed to look like streets and racetrack parts. You can see these parts in Figure 5-31.

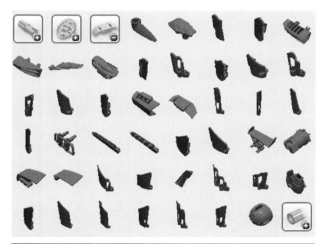

Figure 5-29 Find fenders and other parts of a car's body here.

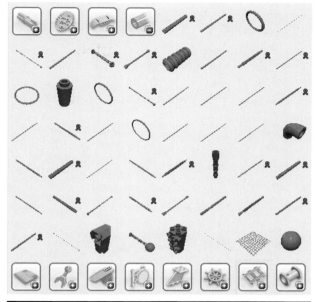

Figure 5-30 This palette holds flexible parts such as air tubes.

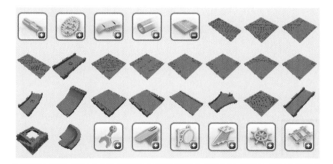

Figure 5-31 Want to make a street? You'll need parts from this palette.

Hero Factory Body Parts

LEGO has a fun product line of giant robots called "Hero Factory." You may also hear the term "Bionicle," which was a previous edition of Hero Factory. LDD has a few palettes of these parts.

Hero Factory parts use a lot of ball joints (Figure 5-32), but the parts also have numerous Technic holes, and most parts therefore can be included in Technic and Mindstorms projects.

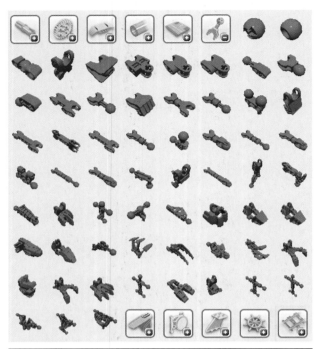

Figure 5-32 Hero Factory body parts are found here.

Hero Factory Accessories

Hero Factory is a line of figures—albeit big ones—so this means the figures have to carry things. This palette (Figure 5-33) has a ton of very colorful and "comic booky" swords, masks, pieces of armor, and even body parts such as tails.

Outdoor Furnishings

Tree parts, flower parts, fences, and a surprising variety of ladders may be found in this palette, not to mention barrels, boxes, pillar finials, and other outdoor accessories. You can see these parts in Figure 5-34.

Vehicle Parts

This palette is a car-builder's paradise (Figure 5-35). It holds much of what you need to build a hotrod or speedboat, including steering wheels,

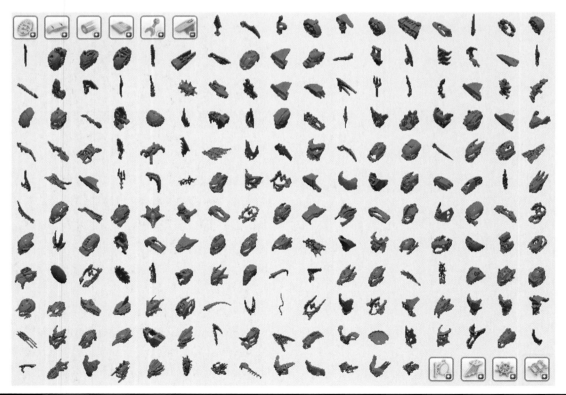

Figure 5-33 Everything your giant robot needs.

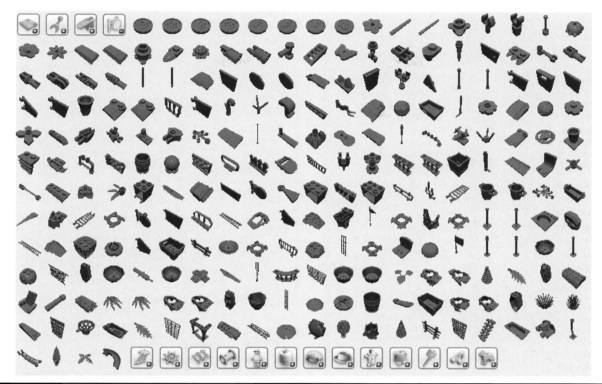

Figure 5-34 This palette holds outdoor furnishings.

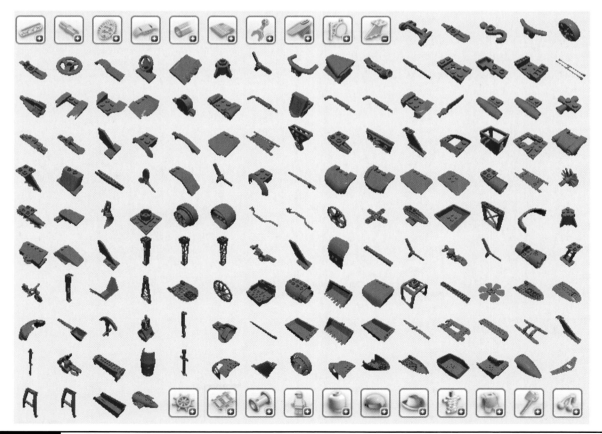

Figure 5-35 Build the race car of your dreams with these parts.

propellers, rocket jets, turbines, mudguards, and so on. There are even a couple of one-piece boats like Zodiacs. Of particular interest to me are the bulldozer shovels, of which there are many (seen toward the bottom of the figure).

Boat Parts

Continuing on with the theme of nautical parts found in the preceding palette, this small palette (Figure 5-36) consists of a number of boat parts: hull sections, masts, an anchor, a couple of propellers, and so on.

Train Parts

Similar to the Boat Parts palette, this one (Figure 5-37) holds train parts such as track sections, wheels, carriage bases, and even a cowcatcher.

Mechanical Parts

This small and seemingly random palette holds miscellaneous mechanical parts such as pulleys and hinges. You can see these parts in Figure 5-38.

Figure 5-37   Build a train with these parts.

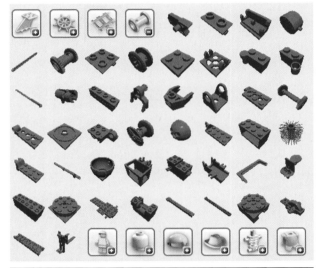

Figure 5-38   This palette holds obscure mechanical parts.

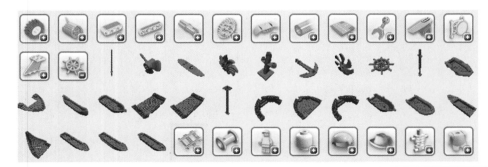

Figure 5-36   This palette offers additional boat parts.

Minifig Torsos and Legs

Now we're getting to some fun stuff: several palettes relating to everyone's favorite, the LEGO minifigure, also known as "minifig." This first of the Minifig palettes holds torsos and legs, some of which have screen printing on them. In addition to variously colored parts, the palette has specialty parts such as a pirate's pegleg or a dwarf's stumpy legs. In order to show the palette's screen-printed parts, Figure 5-39 presumes that you're viewing it in basic LDD. I'm showing most of the other Minifig palettes the same way.

Minifig Heads

The mind boggles at the sheer variety of minifig heads to be found in this palette (Figure 5-40). If you can't find a minifig head here, it may not exist!

Minifig Hair

Keeping the minifig theme going, this palette (Figure 5-41) contains LEGO's library of minifig hair.

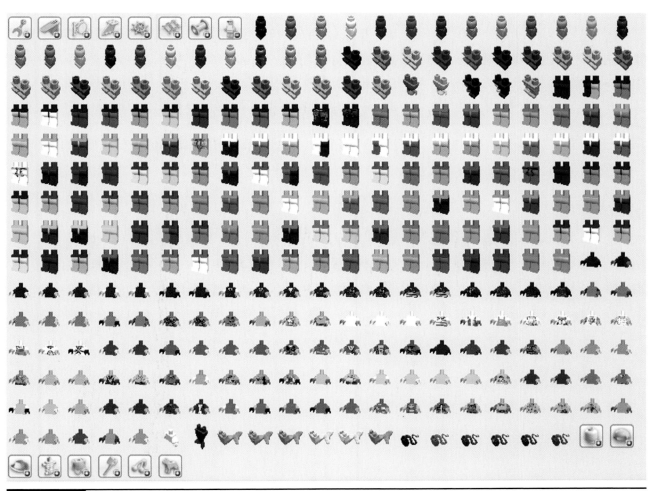

Figure 5-39 Choose your minifig's torso and legs.

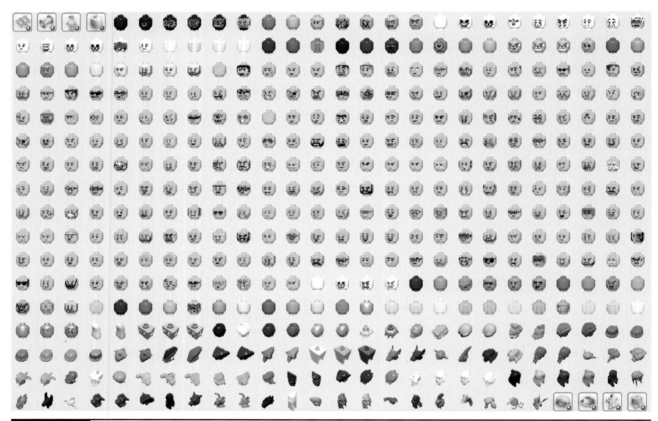

Figure 5-40 Minifig heads galore. No minifig is complete without one!

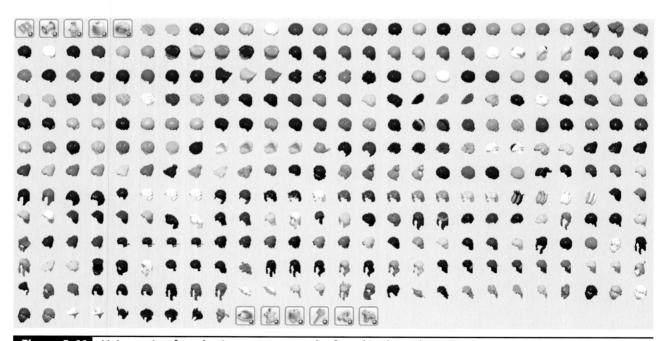

Figure 5-41 Hair ranging from boring to crazy may be found in this palette.

Minifig Hats

Hats, hats, hats. My mind boggles at the sheer variety of LEGO headgear. Figure 5-42 shows all the different configurations.

Alternate Figure Parts

This brief palette (Figure 5-43) holds alternate body parts for skeletons, aliens, and other not-quite minfigs.

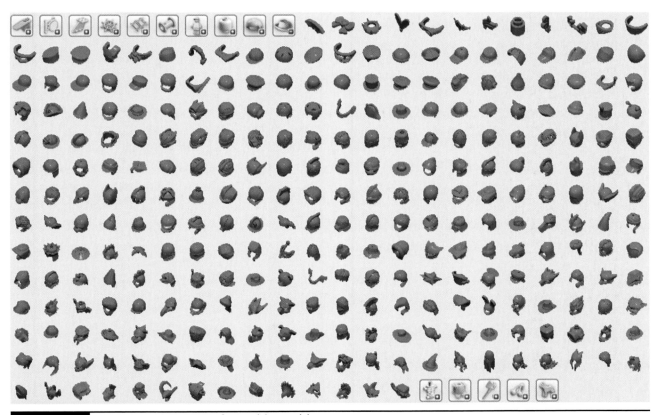

Figure 5-42 More hats than a minifig could possibly wear.

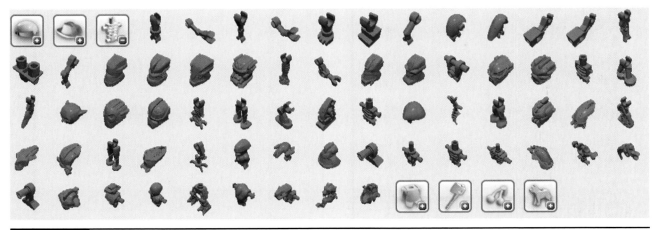

Figure 5-43 Select among a variety of weird body parts.

Minifig Torso Parts

This palette (Figure 5-44) holds accessories for minifigs, including epaulets, neckerchiefs, breastplates, and backpacks. Most of these parts fit over a minifig's neck post and are secured with the figure's head.

Minifig Accessories

This palette offers handheld accessories for minifigs (Figure 5-45) ranging from swords to tools, cups, toolboxes, brooms, crossbows, paint rollers, and countless others.

Food

This palette holds a surprisingly modest assortment of food elements (Figure 5-46).

Animals

The final palette is a great one: animals, glorious animals (Figure 5-47) ranging from small animals such as bats, rats, and kittens to medium-sized ones such as pigs and horses. There is also a selection of body parts from larger creatures such as dinosaur legs, dragon wings, and so on.

Mecabricks

As you might expect, Mecabricks' parts are arranged in a reasonably logical fashion—certainly more so than LDD does, with its occasional randomness. That said, Mecabricks does have its eccentricities. I'll go through each palette and describe what each one holds, and if it's a group of palettes, I'll mention what each individual palette holds.

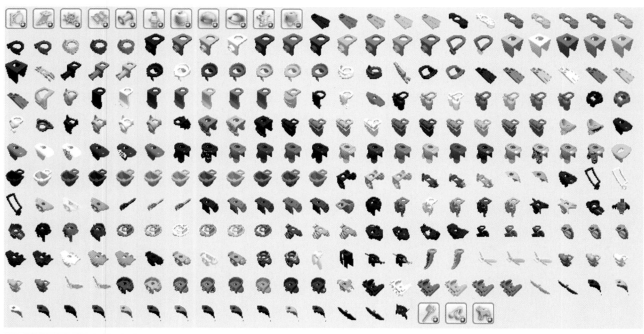

Figure 5-44   Spice up your minifig with a backpack.

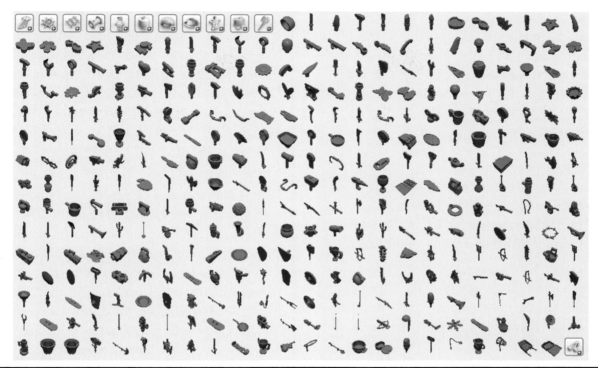

Figure 5-45 Equip your minifig with a hammer, a glass of wine, or an electric guitar.

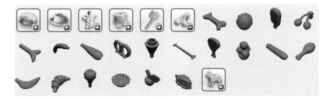

Figure 5-46 Drumsticks, giant pretzels, and bananas typify the sorts of parts found in this palette.

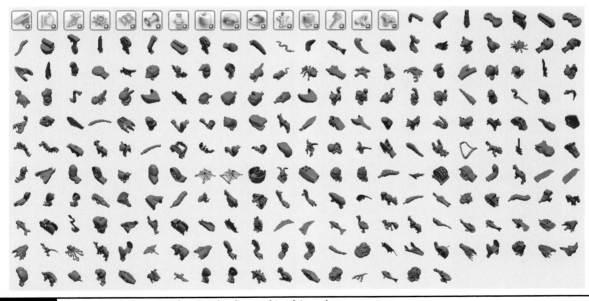

Figure 5-47 Animals and parts of animals abound in this palette.

One caveat I should mention in evaluating Mecabricks in particular is that it is a new application and still evolving. It already compares well with the competition. Therefore, take any complaints I share with a grain of salt.

Finding the Right Brick

Before you delve into the palettes, however, you should learn about the various fields of the convenient table (Figure 5-48) in which Mecabricks describes each part. In this way, you'll have an easy time finding your part.

- **Preview.** A small representation of a brick, much like a thumbnail in LDD. If there is a plus (+) next to it, this means that there are screen-printed versions of that brick available for use.

- **Name and part number.** This is a surprisingly helpful tool because of the sheer number of parts that have been created for the LEGO

System, some of which are deprecated, some out of production, and confusingly, some duplicates of others. Knowing the part number of the element (which is also embossed on the physical part) as well as the part's name helps you to track down the right one.

By clicking on the gear to the right of this field, you'll have the option of seeing this information in English (the default), French, or "Bricklink." The latter is simply the part name corresponding with Bricklink nomenclature. There is a degree of confusion over some bricks' proper names, and many individuals and groups (such as the LEGO fan site Bricklink) come up with their own terminology. By clicking on "Bricklink," you'll simply get slightly different information on each brick, corresponding with the standards on that site.

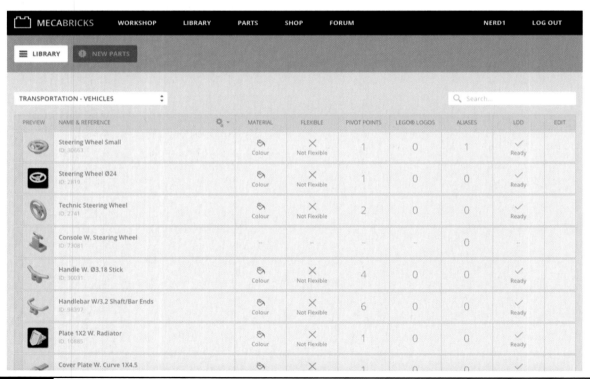

Figure 5-48   Mecabricks' convenient tables organize all the parts.

■ **Material.** Mecabricks divides each part according to the material of which it is made. Plastic and rubber are labeled as "Colour," and anything that is neither gets "Decoration." I can see this category being super helpful, but it isn't right now. For instance, being able to differentiate between rubber parts and plastic parts seems like a natural.

■ **Flexible.** Certain parts can be made to flex in Mecabricks. Typically, these include rubber tubes and belts designed to be strung between two points. These elements get a checkmark in this category. However, not all parts that are in fact flexible in real life fall under this category. The reason is that not every part has been made to work with Mecabricks.

■ **Pivot points.** This field keeps track of all the various ways a part can be rotated, assuming that the part is connected to another part using standard LEGO means.

■ **LEGO logos.** Every stud of every brick has the logo, and there are a bunch of others besides. This field lets you know how many logos the brick has, to further help you to identify the part.

■ **Aliases.** I mentioned earlier, there are some duplicates, and this field keeps track of identical parts with differing part numbers. It's a complicated system, and sometimes these things happen!

■ **LDD ready.** Can you export this model with the expectation that it will work in LDD? You can if all the parts in your model are checked LDD ready.

■ **Edit.** I'm not sure what this field means, and I haven't found a brick that has anything in this field.

The Palettes

The first thing you need to know about Mecabricks' parts palettes is that, confusingly, Mecabricks doesn't call them palettes. They're the "Library," which puzzles me because the "Library" button on the home page takes you to the gallery of models, and the "Parts" button takes you here. I'm going to keep calling them "palettes." And so, without further delay, here are the palettes.

Animals and Accessories for Animals

LEGO offers farm animals, knights' steeds, parrots, monkeys, and a bunch of other designs, and Mecabricks keeps them in this palette, visible in Figure 5-49.

Beams (Six)

Technic beams, used in LEGO's robotics sets, find a home in this collection of six palettes. The first palette may be seen in Figure 5-50.

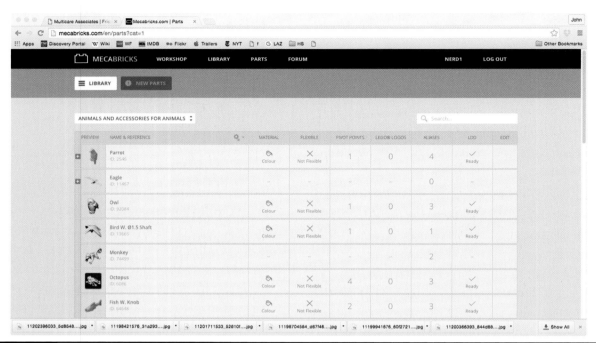

Figure 5-49   Animal parts galore may be found in this palette.

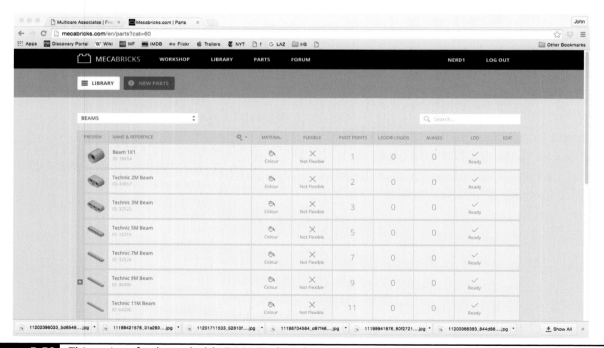

Figure 5-50   This series of palettes hold LEGO's Technic beams.

Bricks (Nine)

LEGO's core part has always been the rectilinear brick, and Mecabricks gathers them together in the Bricks palettes, the first of which is visible in Figure 5-51. Of course, LEGO offers countless variants of the classic brick, so if you're looking for something relatively unusual, you may have to search a bit.

Connectors

Technic beams connect to each other with pegs, a variety of which (Figure 5-52) may be found in this palette.

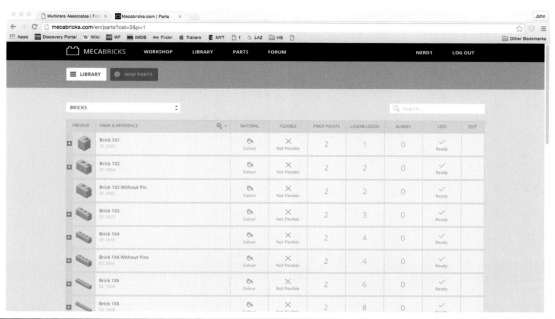

Figure 5-51   Mecabricks' Bricks palettes contain classic and not-so-classic brick-shaped elements.

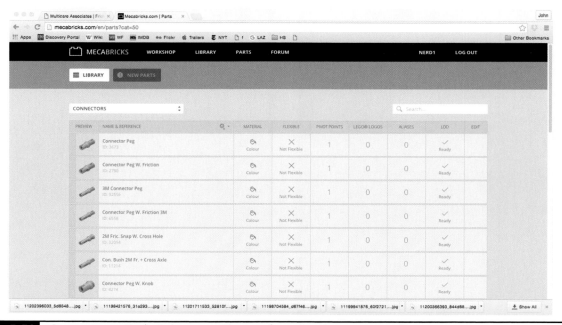

Figure 5-52   Connect your Technic beams with pegs from the Connectors palette.

Constraction

This (seemingly misspelled) palette contains only two weird things in it: a big hand and half a spiked ball. I'm not sure what the story is here. You can see the palette in Figure 5-53.

Cranes and Scaffolding

As one would expect, this palette contains LEGO's crane parts (Figure 5-54) including lattices, scaffold segments, and that sort of thing.

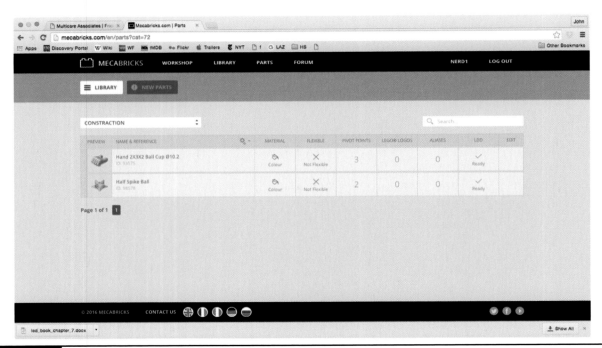

Figure 5-53 This unusual palette holds just two items.

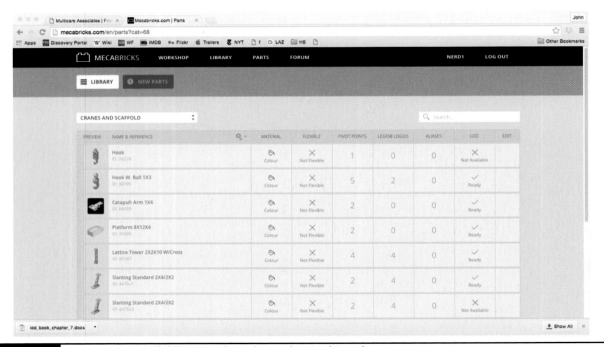

Figure 5-54 Cranes and scaffolds are gathered together in this palette.

Decoration Elements

This vaguely titled palette (Figure 5-55) holds a wide variety of parts, most of which could be described as decoration for a house's exterior.

Doors and Windows

This aptly titled palette holds window and their casements as well as doors and their jams. Figure 5-56 shows the selection.

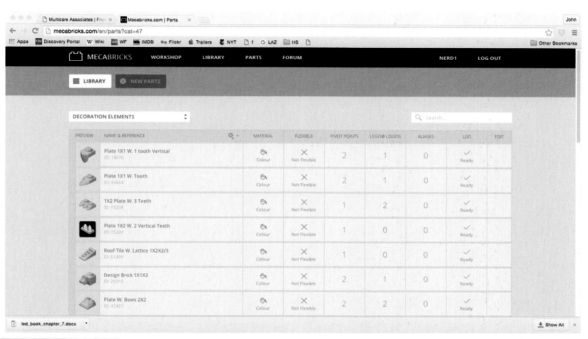

Figure 5-55 Decorate your LEGO house with the help of this palette.

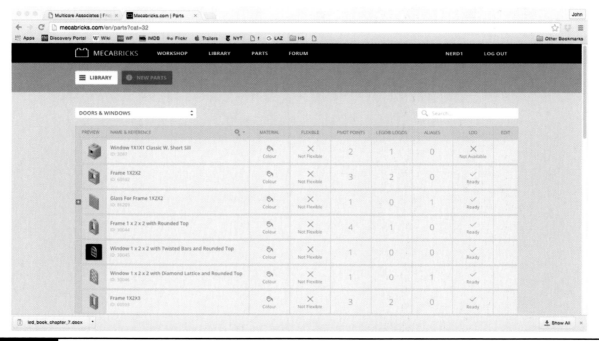

Figure 5-56 A wide variety of doors and windows may be found in this palette.

Electric Parts

This odd palette (Figure 5-57) contains only four elements. Two of these are part of the Technic product line, whereas one of the others is a pushbutton light-emitting diode (LED). The final piece is a light-bulb cover with no electrical functionality on its own.

Fences and Ladders

Another "duh" palette (Figure 5-58). Need fences, ladders, or similar parts? Find them here.

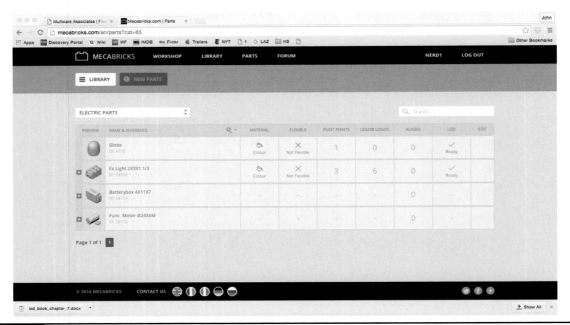

Figure 5-57   This palette contains a small assortment of electrical parts.

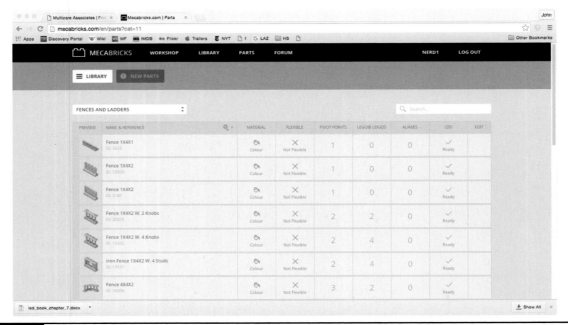

Figure 5-58   Fences and ladders and other elements that resemble them.

Food Stuff

Apples, bananas, and other minifig-sized food items can be found in this palette. You can see a sampling of the menu in Figure 5-59.

Functionnal Elements (Two)

This misspelled set of two palettes hold miscellaneous elements. The first one contains items that resemble parts that actually do a job (e.g., a winch) or simulate them. The second palette (Figure 5-60) holds gears of various types.

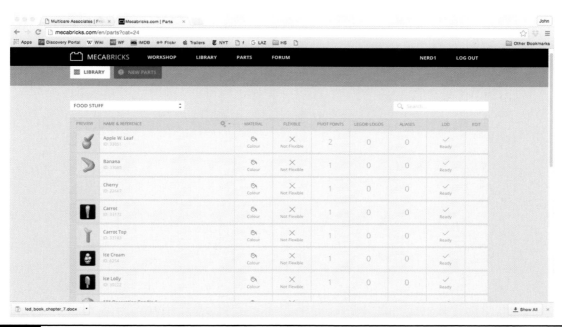

Figure 5-59   A veritable cornucopia of minifig-scale food items.

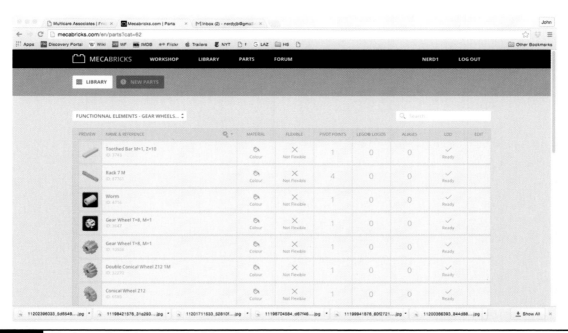

Figure 5-60   Find your favorite gears in this palette.

Interior

Decorate your LEGO house with furnishings from the Interior palette (Figure 5-61), such as chairs, faucet taps, barrels, tubs, and dustbins.

Minifigs (Nine)

Unsurprisingly, the ever-popular LEGO minifig finds itself featured in nine palettes, with one simply for the various heads from which you can choose. The main folder (Figure 5-62) contains some commonplace basics, but the other palettes delve deeply into all the screen-printed variants available.

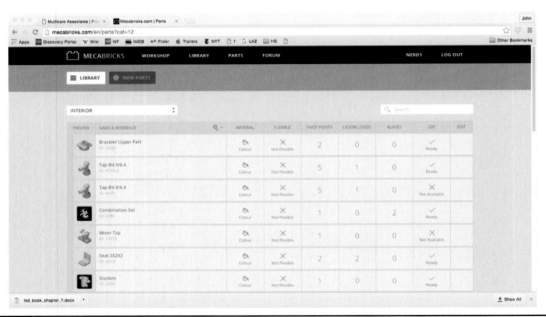

Figure 5-61 Home furnishings scaled for a minifig.

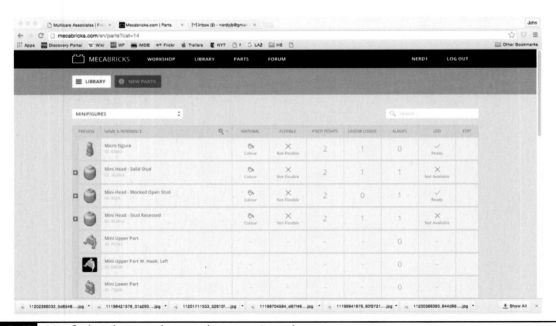

Figure 5-62 Minifig heads, torso, legs, and accessories galore.

Plants

A variety of flowers, trees, and bushes (Figure 5-63) keeps your LEGO garden blooming.

Plates (Six)

LEGO's assortment of base plates (Figure 5-64), angled plates, round plates, and so on may be found in this set of six palettes.

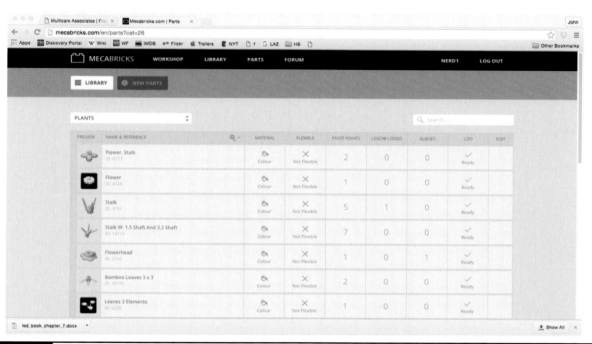

Figure 5-63 This palette gathers together all of LEGO's plant elements.

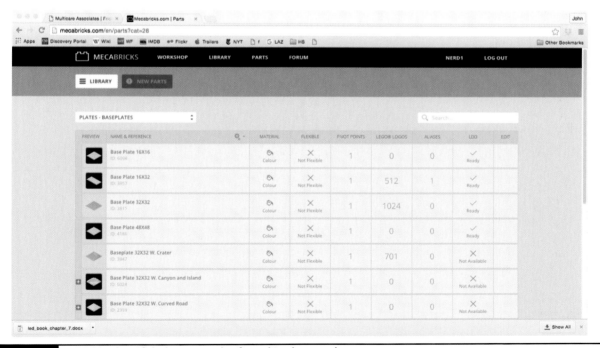

Figure 5-64 LEGO's plentiful plates may be found in these palettes.

Rubbers and Strings

This category includes rubber bands, belts, and strings (Figure 5-65).

Signs and Flags

In addition to the signs and flags themselves, this palette (Figure 5-66) includes flagpoles as well as similar elements such as radio aerials.

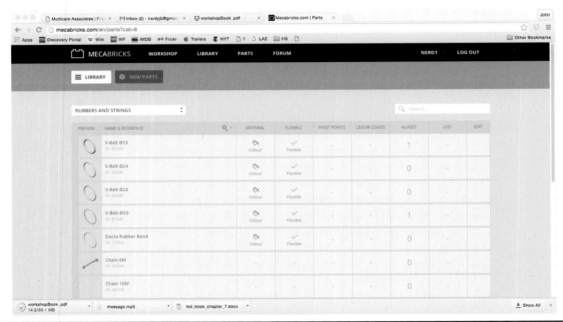

Figure 5-65 Rubber elements are found in this palette, along with strings.

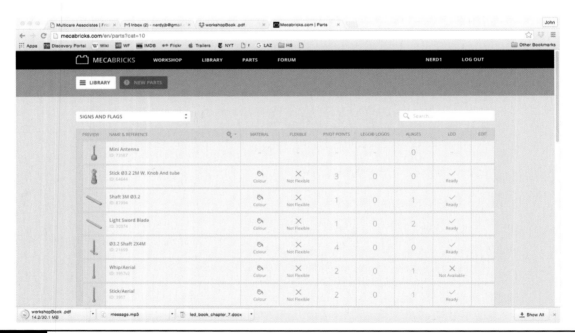

Figure 5-66 Signs, flags, and flagpoles are found here.

Stickers

Instead of holding plastic parts, this palette contains sticker sheets (Figure 5-67) from a handful of sets.

Support

This odd-duck palette holds only four elements (Figure 5-68) and looks as though it could be done away with and its contents placed elsewhere.

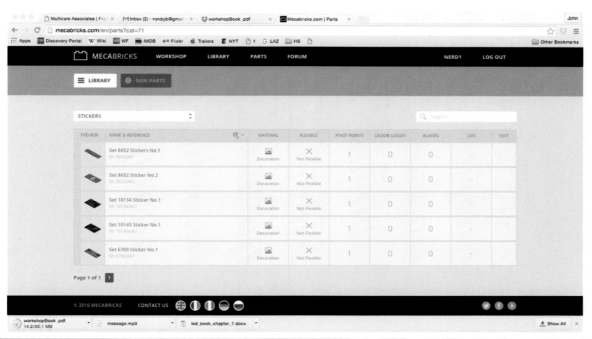

Figure 5-67 Sticker sheets, preserved for future use.

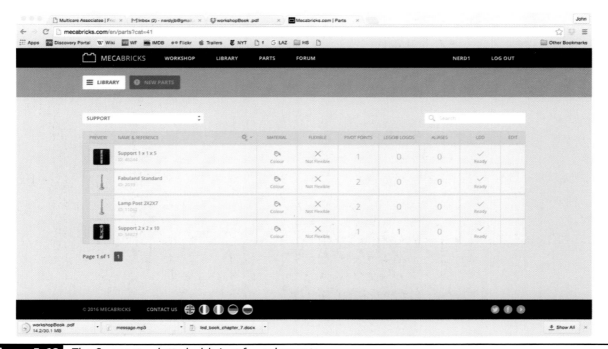

Figure 5-68 The Support palette holds just four elements.

Textiles

You can find cloth elements such as minifig capes, pirate sails, and so on in this palette (Figure 5-69).

Transportation (Four Subcategories)

Mecabricks offers four transportation-related palettes, including airplane parts (Figure 5-70), boats, trains, and cars.

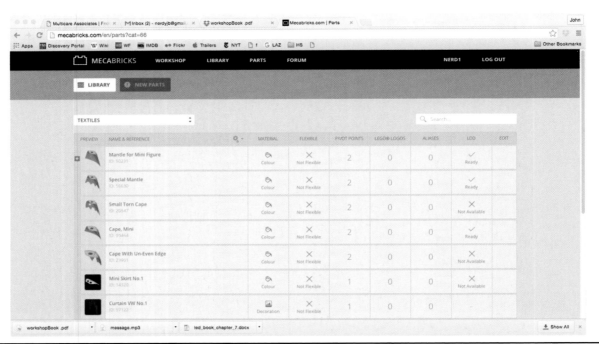

Figure 5-69   Cloth items are found in this palette.

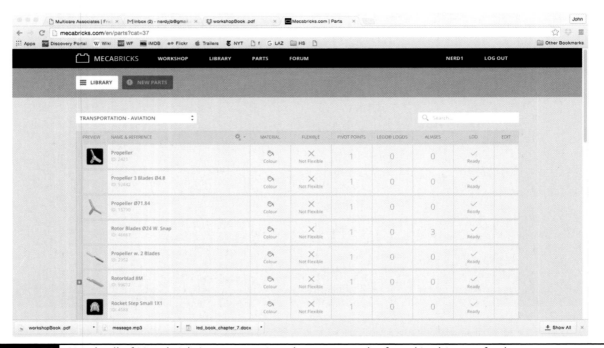

Figure 5-70   Nearly all of Mecabricks' transportation elements may be found in this set of palettes.

Tubes

This is Mecabricks' catchall term for the variety of plastic and rubber tubes found in the set. Figure 5-71 shows some of the options, and you can see how the site breaks down the elements into subcategories.

Tyres and Rims (Three Subcategories)

Mecabricks' creator hails from New Zealand, so he uses British spelling in this series of palettes, one of which can be seen in Figure 5-72. Two of the palettes collect a pair of commonplace rim diameters, whereas the third one holds unusual sizes and miscellanea.

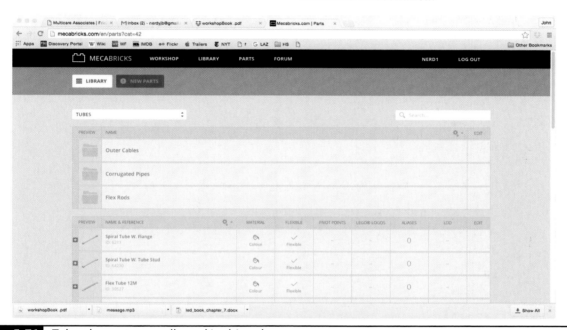

Figure 5-71   Tube elements are collected in this palette.

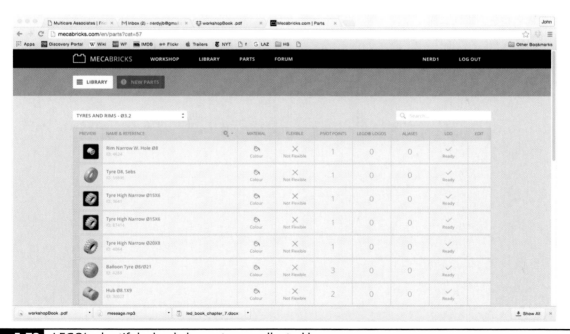

Figure 5-72   LEGO's plentiful wheel elements are collected here.

Various Parts

A truism discovered by those who try to categorize and classify LEGO: *you can't!* Sooner or later you end up with a folder full of random stuff. So it is with Mecabricks. If you can't find it elsewhere, look in the Various Parts palette, shown in Figure 5-73.

Wheel Based

This odd-duck category consists of motorcycle frame elements (Figure 5-74). Not only is it confusingly named (see the next palette), but it is off on its own and not placed with Transportation, as one would expect.

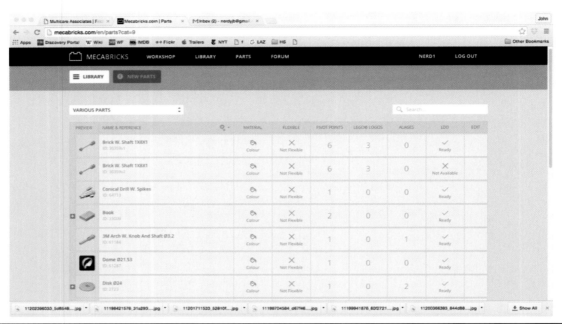

Figure 5-73   A catch-all category of miscellaneous parts.

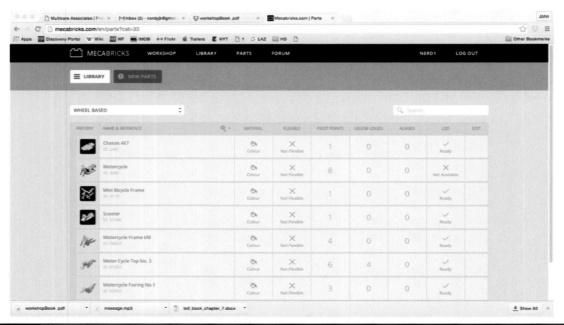

Figure 5-74   Motorcycle frames, stored here instead of in the Transportation palettes.

Wheel Bases

Another weird one, this palette includes small plates with axles protruding from them (Figure 5-75), which raises the same questions as the preceding palette.

Windscreens and Cockpits

This final palette has a variety of curved windows for use in fighter jet models (Figure 5-76).

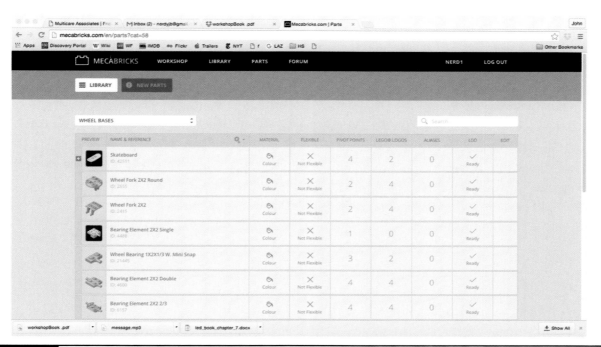

Figure 5-75 This palette contains small plates with wheel hubs attached.

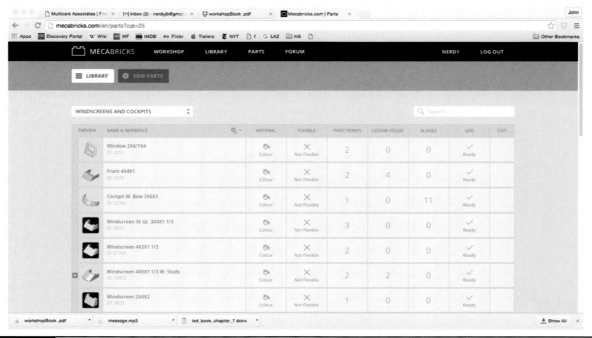

Figure 5-76 The Windscreens and Cockpits palette holds jet plane canopies.

LDraw

Both LDD and Mecabricks, to an extent, organize their bricks in broad categories. By contrast, LDraw has over 120 categories, some containing just a single element or a small group of them. For instance, LDraw has a palette just for car exhaust pipes. Figure 5-77 shows the interface, with a huge pane of palettes on the left, with the right-hand pane showing all the parts in the selected palette and the bottom pane giving you a rendering of the part.

There are both advantages and disadvantages to this setup. The palettes are organized alphabetically, with no attempt to group like parts together. So you'd have to scroll around to find related parts (e.g., boat elements) rather than finding them all together.

Unusual Elements

Instead of covering all these palettes individually (in many cases, they aren't all that different from similar palettes in Mecabricks and LDD), instead, I want to call out just the unusual parts.

LDraw boasts over 7,000 elements, many of which are completely out of production and unlikely ever to return to LEGO's product line. Some of the categories collect the elements from a particular set or product line. For instance, Figure 5-78 shows a raccoon figure from LEGO's Fabuland line, which was introduced in 1979 and lasted barely a decade. But LDraw volunteers digitized those old elements so that you can use them.

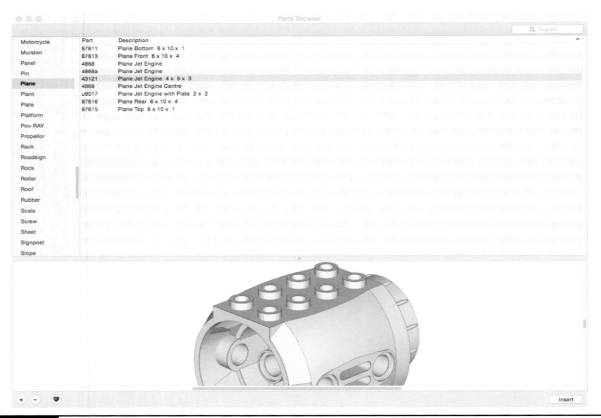

Figure 5-77   Select your parts through this interface.

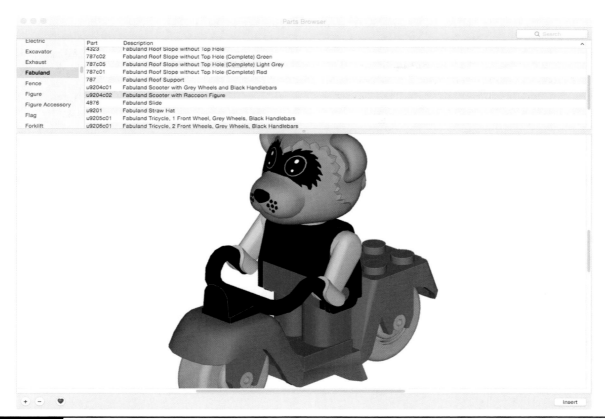

Figure 5-78 LEGO's Fabuland parts are collected for posterity.

LDraw also has an exhaustive collection of parts most grownup builders simply don't care about. For instance, if you want to make a virtual model out of Duplo bricks (Figure 5-79), you can. This comes from LDraw's genesis as a fan-driven project. When someone is passionate to digitize all the Duplo parts, you don't tell that person no. Never mind that only toddlers play with Duplo bricks, which are twice as big as regular LEGOs in every dimension.

Another example of LDraw's legendarily exhaustive library is the Electric palette (Figure 5-80). It contains practically every single battery pack, controller, motor, and sensor found in LEGO's multitudinous product lines. Compare this with just the four found in Mecabricks' Electric palette.

LDraw's Stickers palette is another grouping that shows off the effort LDraw fans have put in over the years. Not just a few sticker sheets,

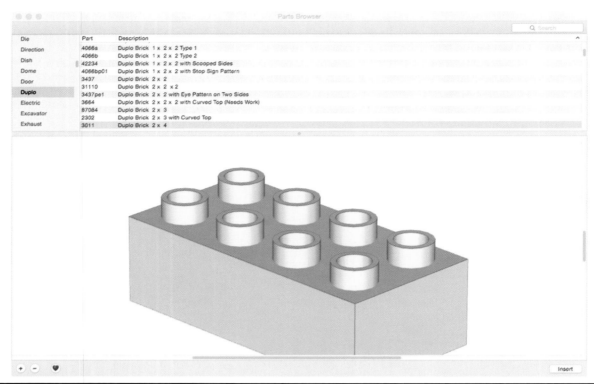

Figure 5-79 Care to build in Duplo, the preschooler's brick of choice?

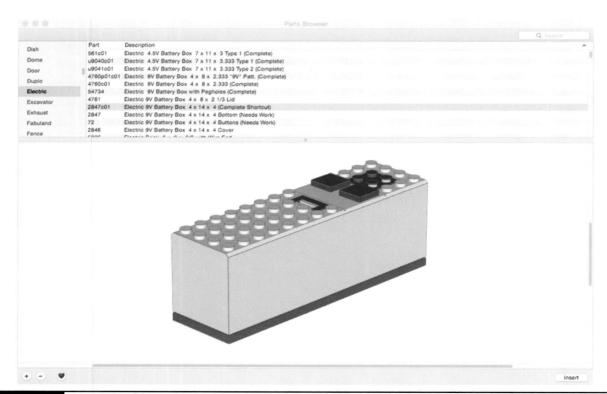

Figure 5-80 This palette offers nearly every electrical part found in LEGO's product lines.

Figure 5-81   Put a sticker on your next LEGO model.

like Mecabricks (or nothing at all like LDD), LDraw fans have digitized dozens and dozens of individual stickers for use in LDraw (Figure 5-81).

I've been enthusing, but some of LDraw's palettes are too big. Take the Technic category. Not only does it have every single part used in the popular Technic line of giant trucks and powered cranes, but it also has obscure parts displayed right next to them. Figure 5-82 shows Technic Figures, a brief and unsuccessful attempt to make human figures for Technic models. They debuted in 1993 and never really caught on. Why should these figures be found in the same palette as Technic beams, which are wildly popular?

Summary

After reading this chapter, you may be sick of LEGO parts, but hopefully instead you are left with the sense that there are nearly limitless possibilities to be had among the brick palettes of LEGO Digital Designer. In Chapter 6, I'll go through the remainder of the interface: the menus that control the LDD experience.

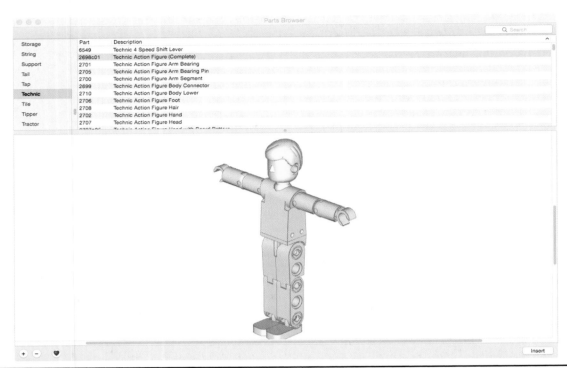

Figure 5-82 Technic Figures, an obscure effort from 1993, may be found alongside more popular parts.

Building with LEGO Digital Designer

NOW THAT YOU'VE DOWNLOADED and installed LEGO Digital Designer (LDD), it's time to get to work! This chapter guides you through the interface and takes you through a few simple building steps.

As is typical with many programs, the various functions of LDD may be accessed through a series of menus found along the top of the home screen (Figure 6-1) and which also may be activated with hot keys. In this chapter I'll explain the various menu items, and I will list the hot-key commands for Windows and Mac. Then I'll guide you through the program's workspace, where all the magic happens.

Figure 6-1 The program opens to a home screen with the regular app grayed out behind it.

Exploring the Application

When the program completes its launch, you're presented with the main LDD screen grayed out, with the brick palettes on the left and the workspace visible on the right in Figure 6-2. Above it are located the usual row of icons.

In front of the grayed-out workspace there is a pop-up window consisting of three big tabs on the top, as well as space for icons in the rectangle below the race car and the white sidebar to the right of the car. The white sidebar has "new" and "open" icons, and the remaining space is filled up with recent programs. If you've just started the program and have never used it, this sidebar will be blank. The space below the image is used for the icons of official Mindstorms

models that LEGO wants to show off. You can always get back to this home screen by hitting the "house" icon in the upper left-hand corner of the window.

Choose the Theme

The three big tabs invite you to choose one of three different flavors of the LDD experience.

LEGO Digital Designer

The blue tab (Figure 6-3), vaguely labeled "LEGO Digital Designer," can be considered a "basic" version. Some options are disabled, and you have an abbreviated parts assortment from which to choose. Not every brick is there,

Figure 6-2   See recent projects and special promotional models on the right-hand side of the home page's main panel.

Figure 6-3   Click on the blue tab for a default LDD experience.

much less in the color you want. However, it has elements the other two options do not. For instance, it features a swath of screen-printed bricks such as minifigure heads and control panels, which you won't find in the Extended palettes.

LEGO Mindstorms

The gray tab, labeled "LEGO Mindstorms," also has a limited parts array and no color changes permitted. The bricks in the palette (mostly) conform to those found in the various Mindstorms robotics sets sold by LEGO, allowing you to build only for that system if you prefer. Figure 6-4 shows a model built in the LEGO Mindstorms tab.

LDD Extended

The black tab is my favorite, labeled "LDD Extended"—this means any brick, any color (Figure 6-5). Rather than showing off every single possibility, the palettes offer one of everything, but all in the same red color, requiring builders to recolor the bricks as needed. While almost overwhelming in its possibilities, the Extended palette doesn't have everything. For instance, you won't find any screen-printed bricks in this palette, although there is a workaround I'll show you later.

Figure 6-4   The gray tab offers a more Mindstorms-focused LDD experience.

Figure 6-5   The black tab selects LDD Extended.

The Three Modes

LDD always operates under one of three modes, each of which offers a unique way to display a model. The program defaults to Build Mode, so many casual LDD users never notice that there are two other modes that might offer a lot to be explored.

The following are the three modes:

■ **Build Mode.** This is pretty obvious: the mode in which you build models. The application defaults to this mode, so when you first launch LDD, it's always ready for you to build.

■ **Building Guide Mode.** This mode is all about sharing your model with others. This mode is discussed at length in Chapter 9, so I won't go into too much detail here.

■ **View Mode.** This mode is for displaying your model attractively, with a background image rather than the ugly gray workspace. Figure 6-6 shows an example of View Mode, with a colored background substituted.

The icons in the upper-left corner guide you to three main things to do with the View Mode:

■ Pressing the "camera" icon takes a screen shot. LDD screen shots are PNGs with transparent backgrounds, making them web ready as well as easy to use.

■ The icon with the movie reel "explodes" the model and makes it seem to fly to pieces. It's fun and cinematic, but that's about all.

■ The "globe" icon triggers the background images. There are four that default with the program—space, a mountainside, a sea, and a surreal golden desert. You also get a looped sound effect to go with each background.

One downside to View Mode is that you can't make changes to the model. You must switch

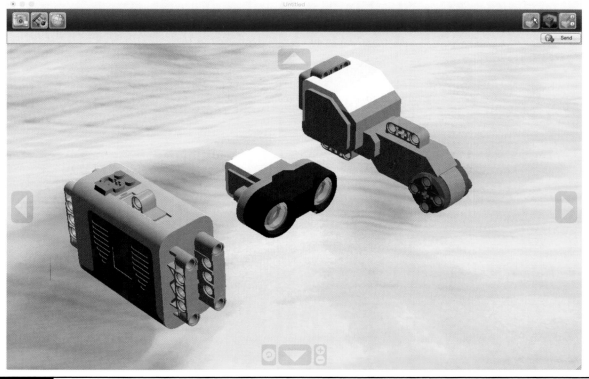

Figure 6-6   View Mode helps you to display your model.

back to Build Mode to make any edits. This can be a good thing, however, if you want to show off a finished model!

Menus

Convenient menus may be found at the top of the screen, and if you're not a hot-key user, you're likely to interact with these menus quite a bit. Let's go over each item in turn.

Preferences

This menu contains a series of options that allow you to turn on and off LDD features. You can find it in the Apple menu on Macs and in the main LDD menu on PCs.

Show Information Field

This makes use of the gray bar at the bottom of the workspace to show information about a part selected. This is great for sussing out the right parts for your model.

Show Tool Tips

This function controls those little pop-up windows that explain what the buttons do when you put your mouse pointer on them.

Enable Sounds in the Application

LDD makes clicks and chirps to indicate when various events take place. It's generally quiet and inoffensive, but if you want it to be silent, uncheck this option.

"Keys for Turning" Shown Along with Cursor

When you are "holding" a brick or assembly of bricks with your mouse pointer, you have the option to rotate it, and checking this option causes an icon to appear on top of the element to let you know that it can be rotated.

Repeat Inserting Selected Bricks

This option allows you to place many instances of the same brick very quickly.

Show/Hide Brick Count

The information bar at the base of the workspace also may be configured to show the total number of LEGO elements in the project.

Invert Camera X-Axis

This setting allows you to reverse the way you rotate around the model. Sometimes controlling something with ten keys works a different way than you're expecting. For instance, does clicking on the right arrow make the workspace rotate counterclockwise when you're expecting clockwise? Simply tick this setting, and it will work in reverse.

Invert Camera Y-Axis

This is the same as for the X-axis, except that this reverses the controls for the Y-axis.

> *Tip:* The X and Y refer to the axes along which the model rotates, as seen in Figure 6-7. Clicking on a left or right arrow rotates the model horizontally. This is the X-axis. Clicking on the up or down arrows rotates the model vertically so that the far end gets closer and the near end gets further away.

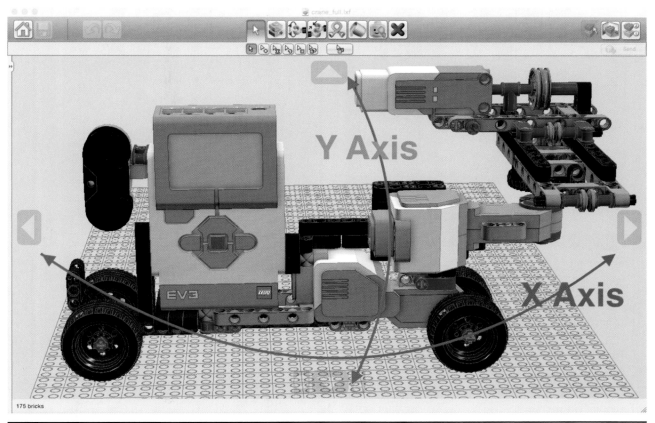

Figure 6-7   Your model can be rotated along two axes.

High-Quality Rendering of Bricks Placed in a Scene

For computers with display problems, LDD can be instructed to render the elements more crudely. To be honest, however, I can't really tell the difference.

High-Quality Rendering of Bricks in the Brick Palette

Similar to the preceding item, this item doesn't seem to make any difference to me.

Outlines on Bricks

Sometimes boundaries between bricks of the same color are lost in the rendering process, and a stack of elements looks like one big one. If this is a problem, this option allows you to mark the edges of the bricks more prominently. Figure 6-8 shows a model with outlines, and Figure 6-9 shows the same model without the outlines. Choose whichever works better for you! Be aware, however, that each facet's outline represents another rendered element, which adds to the memory demands of the model. Depending on your computer, employing this option may very well slow down LDD's functions.

Figure 6-8   The model with outlines added to each part.

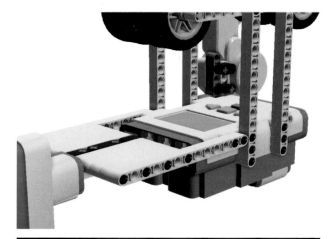

Figure 6-9   The model without the outlines.

Advanced Shading

This is another function that adds another layer of shading to the parts displayed. This amounts to more polygons. I could not really see any difference on my screen.

Compatibility Mode

This launches LDD in "safe mode" if you don't want to or cannot connect to the network.

Reset Preferences

This resets all preferences to their defaults.

> **Tip:** In many instances you'll need to restart LDD to see the changes you just made to your preferences. The program will alert you when this is the case.

File

If you've used computers before, you'll mostly find the contents of the "File" menu to operate as expected. I'll cover each item and also include a table of hot keys (Table 6-1).

Table 6-1	File Hot Keys	
Command	**PC**	**Mac**
New	CTRL+N	CMD+N
Open	CTRL+O	CMD+O
Import Model	CTRL+I	CMD+I
Export Model	CTRL+E	CMD+E
Save	CTRL+S	CMD+S
Save As	SHFT+CTRL+S	SHIFT+CMD+S
Print	CTRL+P	CMD+P
Export BOM	CTRL+B	CMD+B

New

Click on this item to create a new file. If you have unsaved changes in your current project, it will prompt you to save before closing it. Just in case it slipped by, know that you can have only one file open at a time. If you have an open file, and want to begin a new project, you must either save and close the open file or discard your changes.

Open

If you want to open an existing file, this option will first prompt you to save any unsaved changes in your current project, though it will not dump your current file if you elect not to save—only when the new project opens up will the old one be closed, so if you change your mind, you won't lose any unsaved changes.

Import Model

This valuable tool imports an entire LDD file into your current file. This is great if you have parts of a larger project stored as pieces in their own files—simply import each segment, and assemble the full model.

Export Model

This menu item exports the current file in one of four formats: three of them are variants of LEGO's proprietary LXF format, including two variants that use elements of the popular XML markup language. The fourth option allows you to export into the LDraw format. This means that you could drop your model into an LDraw editor if you end up liking it better. Figure 6-10 shows a LDD model brought into the LDraw editor Bricksmith using this technique.

Save

Save your file. This function is grayed out if you have not made any edits to the current file.

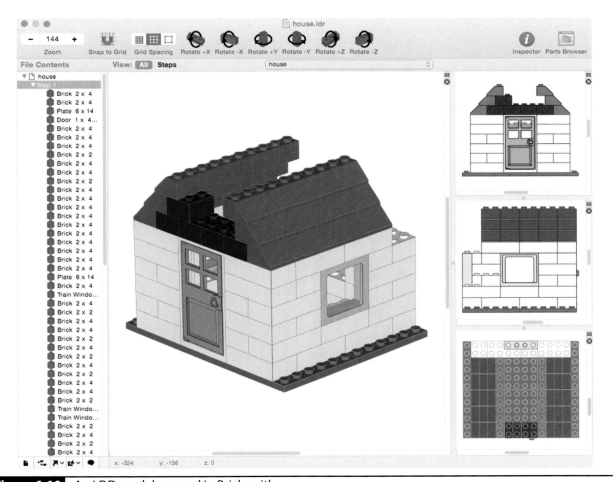

Figure 6-10   An LDD model opened in Bricksmith.

Save As

This menu item works as one would expect, allowing you to save an open file under a different name.

Print

Another easy one. Print all the elements that are in the document, regardless of whether they are visible in the pane. The angle of the print is what you are looking at on the screen, but zoomed in to show every part.

Export BOM

A BOM is a "bill of materials"—a listing of every LEGO part in the LDD file. It's saved in either Microsoft Office Open XML Workbook format (.xlsx), or a ZIP archive of the same file. An easier way to get a BOM involves using the "Building Instructions" function. I'll show you how to do this in Chapter 9.

Edit

This category of menu options includes many of the classic Edit tools, such as Cut and Paste, as well as some LDD-specific options. There is a table of hot keys (Table 6-2).

Table 6-2	Edit Hot Keys	
Command	**PC**	**Mac**
Preferences	CTRL+6	CMD+,
Undo	SHFT+CTRL+Z	SHFT+CMD+Z
Redo	CTRL+Z	CMD+Z
Cut	CTRL+X	CMD+X
Copy	CTRL+C	CMD+C
Paste	CTRL+V	CMD+V
Select All	CTRL+A	CMD+A
Group	CTRL+G	CMD+G
Save to Template	CTRL+ALT+G	CMD+ALT+G

Preferences (PC only)

Application preferences are located in Edit on PCs and in the main LDD menu on Macs. See the "Preferences" item at the beginning of this chapter to see what options lie within.

Undo

The usual Undo feature. This option may be selected only when there are unsaved edits to the project. If you haven't made any edits, the icon on the workspace pane will be grayed out. Note that Undo only undoes actions within LDD, not other applications or the desktop.

Redo

This menu item redoes a command you canceled with an Undo. As with Undo, if there are no commands available to Redo, the option is grayed out.

Cut

This is the standard Cut function, where selected elements are deleted off the workspace but remain in memory. Note, however, one critical difference: the shapes you cut cannot be pasted into any other program, and moreover, cutting an element doesn't replace whatever you have stored on the clipboard from another program.

Copy

Copy stores a copy of the selected bricks on LDD's internal clipboard while not deleting those elements in the workspace. As with Cut, you don't replace the computer's main clipboard contents when you Copy an element.

Paste

This function takes bricks saved on LDD's internal clipboard and pastes them onto the workspace. If nothing is contained on the clipboard, then this option is grayed out. Note that any non-LDD data on the clipboard will not be affected.

Delete

Any LDD element selected when this option is hit will vanish from the workspace, and you'll need an Undo to get it back.

Select All

This option selects every element in the workspace, whether visible or not.

Group

This command behaves differently depending on what you're doing. Basically, it defaults to whatever the program *thinks* you want to do. If you're selecting multiple elements when you hit Group, it will create a new group from the selection. If you're selecting one or more objects that are already in a group, triggering this command splits them off into a subgroup. I'll show you how to use the Group function in your models later in this chapter.

Save to Template

This command creates a new template from the selection. I'll talk more about using templates as part of your work later in this chapter.

Toolbox

The tools found here duplicate those in the Menu bar, and you can access them either through the text menus, hot keys (Table 6-3), or by clicking on the icons above the workspace.

Table 6-3 Toolbox Hot Keys		
Command	**PC**	**Mac**
Select	V	V
Toggle Select tools	SHFT+V	SHFT+V
Hinge	H	H
Hinge Align	SHFT+H	SHFT+H
Clone	C	C
Paint	B	B
Hide	L	L
Delete	D	D
Take a Screen Shot	CTRL+K	CMD+K
Generate Building Guide	CTRL+M	CMD+M

Selection Tools

These tools offer ways to select one or more bricks. As your builds become more complicated, being able to grab specific parts in a giant tangle of elements becomes vital. A fair question might be, what actually does "selection" entail? Later in this chapter I'll describe the two modes of selection and how you use them while building.

Single.   Click on an element in the workspace, and it will be selected. Click on another item, and that one will select while the first one deselects.

Multiple.   Select multiple objects with this mode, clicking each one in turn.

Connected.   This option selects the element clicked as well as every other element connected to the first one.

Color.   When you select an object of a certain color, all other objects of that color become selected as well. Figure 6-11 shows a model with all the black bricks selected.

Shape.   Select that object and every other part of the same shape, for instance, every 2 × 4 brick in the workspace.

Color and Shape.   All elements sharing the same color and shape are all selected.

Invert Selection.   This works as one would expect: select a brick, then click the tool

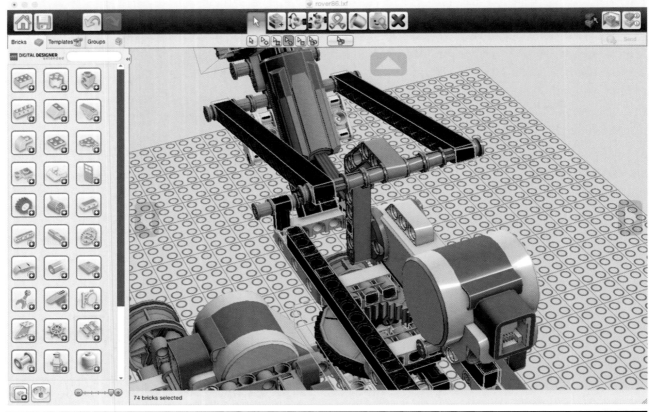

Figure 6-11 All elements of a single color selected.

to deselect that first brick, and then select everything but that brick.

Next Selection. This tool toggles between the various selection tools. When used with hot keys, it lets you find the specific tool you want very quickly.

Clone

When this tool is selected, click on an element to create a duplicate, selected and floating. You can keep clicking and keep cloning and make a huge bunch of parts.

Hinge

This tool helps to manipulate models with moving parts. For instance, a wheel on an axle could be rotated with the help of the Hinge tool.

Hinge Align

Hinge Align helps to move a part so that it fits in with another element. For instance, if you want to connect a beam with another beam, you can click on one mounting hole and then another mounting hole, and the beam rotates to align the holes.

Flex

This tool bends flexible parts. A flex axle, for instance, is a rubber axle that can be bent around a corner to simplify rotating a wheel. This tool helps to connect the axle to its appropriate mounting hole.

Paint Tools

This suite of tools controls the coloring and decoration of bricks in Extended. If you attempt to use this tool in the basic LDD and Mindstorms themes, you'll find the options to change the color and decoration of bricks missing because you can't modify bricks in those themes.

Paint. The default subtool within Paint is called Paint as well and governs changing the color of the bricks.

Color Indicator. This small icon indicates the current color selection.

Color Picker. This little eye-dropper tool sets the Color Indicator to match the color of whatever brick you click on.

Decoration. This tool helps you to decorate a brick with screen printing. I'll show you how to do this later in this chapter.

Hide

Use this tool to make bricks disappear from the screen without them actually moving or getting deleted. Select a brick, and hit Hide, and the brick seems to vanish, but it's still there. You can select it and interact with it, but you can't see it until the parts are unhidden. Hide is often used to get at the interior of a complicated model without altering the model in any way.

Delete

Select this tool to delete any selected parts; then any additional parts you click on will likewise be deleted. Use Undo if you delete the wrong element by mistake.

Take a Screen Shot

Take a screen shot of your workspace screen, cropped as you see it on the screen. The resulting picture has a transparent background and won't include the gray workspace at all. Most of the LDD screen shots in this book employed this technique to generate.

Generate a Building Guide

This command autogenerates a building guide for your project. I'll describe this aspect of the program in greater depth in Chapter 9.

View

This menu governs what you see and can do within LDD. As with the other menus, you can either click on an item or use hot keys (Table 6-4).

Table 6-4 View Hot Keys		
Command	**PC**	**Mac**
Build Mode	F5	F5
View Mode	F6	F6
Building Guide Mode	F7	F7
Send to LEGO.com	SHFT+CTRL+B	SHFT+CMD+B
Show/Hide Camera Control	CTRL+1	CTRL+1
Show/Hide Brick Palette	CTRL+2	CTRL+2

Build Mode

The program defaults to Build Mode, but you can return if you happen to be in View Mode or Building Guide Mode. As mentioned earlier, you do your model creation in Build Mode.

View Mode

This switches LDD into View Mode, which gives you options for displaying your model.

Building Guide Mode

This switches LDD into Building Guide Mode, which guides you through the process of creating assembly instructions for your model. Chapter 9 shows a number of ways to do this, not only for LDD but also for LDraw's solution, LPub.

New Themes

This misnamed menu item lets you switch between basic LDD, Mindstorms, and Extended palettes. If you built a model in one theme, when you use this function it becomes part of the new theme for purposes of brick selection and managing templates.

Send to LEGO.com

Use this feature to send your model to LEGO's Mindstorms page and show it off to the world.

Show/Hide Camera Control

The translucent arrows that rotate the camera around the model can be made to disappear, which means that you would have to use hot keys to move the point of view.

Show/Hide Brick Palette

This command minimizes the brick/group/ template palettes on the left-hand side of the workspace, giving you more room to work.

Help

This section provides modest help for those who need it, though mostly it's just a short manual, downloaded the same time you grabbed LDD and resident on each user's computer.

Search

This function searches throughout the various menus, so if you want to find a specific command, this is the way. However, you can't search on elements in the Help section's search menu.

Help

This launches the help document for the application. It's a browser-accessed HTML file with brief details of every function. It's more of a user's manual than a proper help document, however.

Privacy Policy

In the unlikely event that you want to read LEGO's online privacy policy, here's your opportunity.

Checking Out the Workspace

The home screen defaults to the blue tab. I hesitate to call it the "easy" set, but it definitely has fewer customization options, so maybe easy is somewhat accurate. If you're on the blue tab, then click on "Free Build" to create a new document.

You might want to save this file—go to File > Save As, and name it whatever you want. This file will now show up on your LDD home page's "Recent Files" pane, allowing you to quickly access it in the future.

> *Tip:* Only one LDD file may be open at any time. You can work around this by cutting and pasting, importing, and using templates. I'll explain these techniques in later chapters.

Follow along with Figure 6-12 to learn about the interface and its components.

A. Home

The "Home" button makes the home screen pop up on top of the workspace. This has the effect of closing the present file, and you will be prompted to save it if you've made edits.

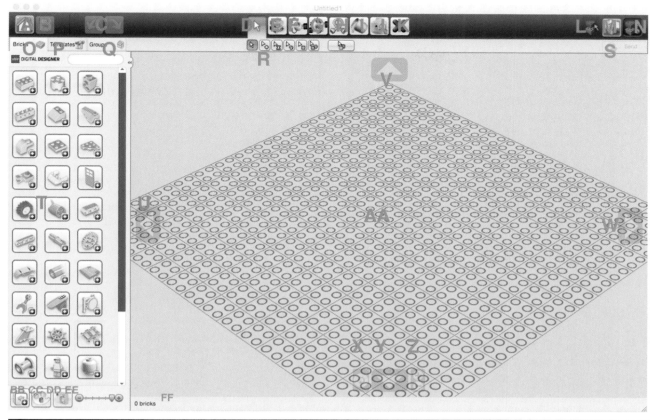

Figure 6-12 The workspace interface and its components.

B. Save

The "Save" button duplicates the typical Save function of any program. If no edits have been made to the design, the icon is grayed out.

C. Undo and Redo

"Undo" and "Redo" buttons become blue if those actions are possible or they are grayed out if they aren't available.

D. Select

The Select tool offers different ways to grab hold of parts.

E. Clone

The Clone tool duplicates parts so that you don't have to look through the palettes to find a commonplace brick—just select one on your desktop, and click on the "tool" icon to dupe it.

F. Hinge

The Hinge tool allows you to manipulate moving parts like, well, hinges—not to mention axles, knobs, and anything that can be moved or rotated in real life.

G. Hinge Align

The Hinge Align tool refines the Hinge tool to rotate to a specific direction, such as to align with another part.

H. Flex Tool

The Flex tool helps you to flex parts that bend. There are not many of them, but this tool is a great help for making these parts fit into your model.

I. Paint Bucket

The Paint Bucket changes bricks' colors in LDD Extended.

J. Hide

The Hide tool hides selected bricks, helping to keep your workspace clear of parts you don't want to delete but don't want cluttering up the place. It's a cinch to unhide the parts when you're ready.

K. Delete

The "Delete" button deletes any bricks selected, as well as any brick you subsequently click on while using Delete.

L. Build Mode

The "Build Mode" button, shown pressed already—because you're building!—toggles the building view.

M. View Mode

The "View Mode" button superimposes your model on a background, making it seem to float in space (for instance) or some other background, and allows you to take screen shots of your creation with the background.

N. Building Guide

"Building Guide Mode" takes you to the mode.

O. Brick Palette

"Brick Palette" selects the library of parts you'll use to make your model.

P. Templates

Templates are user-organized sets of bricks that are available anytime you're working in LDD, regardless of the specific file. I covered these in Chapter 4.

Q. Groups

Groups are sets of bricks within the build. Say I create the perfect car in LDD; I could group those car bricks together into one unit. I talked about groups as well in Chapter 4.

R. Tool Options

The small set of "Tool Options" icons offers options on the tool selected. I covered each tool in Chapter 5 as well as their options.

S. Send

"Send" uploads the model to Lego.com. I'll explain what's involved in Chapter 9.

T. Brick Palettes

Each of the gray dividers in the figure—there are 37 total, including some you have to scroll down for—is a folder full of LEGO parts. Note the Search field in the palette as well.

U. Rotate Left

Rotate the camera left. The "camera" is our viewpoint as we look at a three-dimensional model.

V. Rotate Away

Rotate the camera away.

W. Rotate Right

Rotate the camera right.

X. Undo Camera

Undo camera movement. This doesn't undo something like a brick being moved. Instead, it's a separate Undo just for the camera.

Y. Rotate Back

Rotate the camera back.

Z. Zoom

Zoom controls.

AA. Workspace

Workspace.

BB. Expand Dividers

These are the folders of bricks, and this tool opens up all the dividers so that you can scroll through every brick.

CC. Sort Bricks by Color

In basic LDD, this is useful because not every brick is available in every color.

DD. Filter Bricks by Box

This feature is mostly unrealized and would allow you to show only the bricks contained in a specific LEGO set. However, it has only one filter, which scans for Hero Factory parts. This is a nice idea if it's ever fully implemented!

EE. Thumbnail Size

This tab bar changes the size of both the thumbnails and folder icons in this pane.

Navigating in the Workspace

One of the challenges you'll encounter while working in LDD is the workspace. You're navigating around a three-dimensional object, and this involves rotating around the build area as well as zooming in and out.

Camera Controls

You'll notice four arrows positioned around the screen, showing you the direction the camera travels—as you recall, the "camera" is the view direction, as if you were moving a camera around a real object and looking through the camera's viewfinder. Figure 6-13 shows the Basket Tribot model from the LDD Mindstorms tab. Launch it yourself, and click on the arrows to see how you can move around the model.

- The left arrow rotates the workspace clockwise. You can also hit 4 on the numeric keypad.
- The right arrow rotates the workspace counterclockwise. If you're using the 10-key pad to navigate, click on 6.
- The top arrow rotates the workspace away from you. This equates to 8 on the 10-key pad.
- The bottom arrow rotates the workspace toward you. Hit 2 on the 10-key to get the same effect.

Using the Mouse to Move the Camera

A great way to quickly navigate around a creation is to use the mouse's right button. Click and hold, dragging the mouse around to change the view angle without the hassle of arrows. This also works with the thumbnails in brick palettes, allowing you to look at a part from all angles before grabbing it out of the palette.

Figure 6-13 Basket Tribot model from the LDD Mindstorms tab.

Zooming

The plus (+) and minus (–) are zoom controls. They allow you to come up as close as you need to the model, enough so that an ordinary peg fills the screen. You can also zoom using your mouse if you have the kind with a click wheel in between two buttons.

Reset View

If you're zoomed in or out and want to return the zoom level to the default, click the "Reset View" button to the left of the down arrow. This zooms back to show all the bricks and models on the screen, though it won't change the angle of the creation. You can also reset the view by hitting 5 on the 10-key pad.

Panning

One view option exists only as a hot-key command. Hold down SHIFT, and right-click to pan up, down, right, or left.

Hiding the Palettes and Camera Control

While viewing your creation in Build Mode, you might be tempted to see the robot without the annoying navigation arrows and brick palettes in the way—I know I do! Figure 6-13 shows the Tribot without the clutter.

- To hide the camera controls, click on View > Show/Hide Camera Control. The hot keys are CMD+1 on a Mac and CTRL+1 on a PC.

- To hide the brick palettes, select View > Hide Brick Palette. You can also use hot keys to trigger this change, using CMD+2 if you're on a Mac and CTRL+2 if on a PC.

When both are hidden, the uncluttered display offers a great canvas on which to work.

Building Basics

In this section, you learn how to build a simple model in LDD. To begin with, you must find the element, then add it to the workspace, and then move it around.

Finding an Element

Forget about building for now. To place your first brick, you still have to find it. Here are some ways of finding that perfect brick.

Look for It

Very often the easiest way to find a part consists simply of going into the brick palettes and looking at the thumbnails. You can press CTRL (PC) or CMD (Mac) with a mouse drag to rotate the thumbnail to get a better view of it or simply right-click and drag.

Search

Often you can enter an obvious search term (such as "wheel" or "Technic") and find the right part that way. Also, in LDD (as well as in LDraw and Mecabricks), you can search on brick part numbers, which allows you to track down the precise element. Of course, this presumes that you know the number! Finally, LDD (basic) and LDD Mindstorms allow you to search for bricks by color (Figure 6-14), which reveals all bricks with matching color. LDD Extended, with unlimited choices in brick color, lacks this option.

Figure 6-14   Sort bricks by color to narrow your options.

Research

I often look up parts by researching on LEGO fan sites such as Bricklink.com, which lists every single part and in which set each may be found. You can look up a set of your own to see what parts are in it—and therefore would be found in your parts bin. Find the serial number for any LEGO element simply by looking at it; printed somewhere on the brick is a serial number.

Maybe It Doesn't Matter!

If you're not planning to build the model in the real world, who cares whether you find the right parts? If it looks good, it is good. Stop thinking about LDD as being a simulation of the real world and think of it as a digital art program—everything is on the table.

Adding, Moving, and Placing an Element

Simply click on an element in your brick palette, and the part will appear in your workspace selected, as shown in Figure 6-15. You have a number of options:

- Click and hold to move the brick wherever you want.

- Click and release to place the element just as you see it.

- Use your 10-key pad to rotate the part in 90-degree increments.

If you have more than one part selected, LDD will treat all the parts as one group, and you can maneuver it just as easily as you could a single part.

Colorizing and Decorating an Element

One of the best aspects of virtual building is not being limited by real-world considerations such as whether you actually own that brick in that color! LDD's Paint tool allows you to change a brick's color (with limitations, as I'll describe) or add a printed design to it.

Coloring an Element

Depending on your theme, you have a number of options for coloring a brick. In the "basic" LDD theme as well as in the Mindstorms theme, you may choose among a fixed number of colors in the Bricks palette and cannot change the set.

For those using LDD Extended, however, the sky's the limit! You can make any brick

Figure 6-15   Clicking on a part in the palette places it in the workspace.

any LEGO-approved color, whether opaque or transparent (Figure 6-16). Use one of these techniques to change a part's color:

- Select an element or elements, and then click on the "paint bucket" icon to select the Paint tool; then click on the color swatch (which defaults to black) to choose a color. When you click on a color, the brick(s) you selected immediately change to that color.

- If you're already in Paint, you may click on individual bricks to change their color.

Decorating an Element

Some LEGO parts feature printed designs on them, like the dashboard of a race car or a scientist's computer. You can access these in LDD, and it's pretty easy.

You are limited to the parts ordinarily decorated. For instance, the 2 × 2 slope brick is used for control panels in LEGO models, and you can choose among different types of control panels. By contrast, you couldn't put that kind of design on a minifig head; you are limited to the existing minifig head art.

To use one of these bricks, you simply place the part to be decorated in the workspace, give it whatever color you want, and then select among the images (Figure 6-17) to decorate it.

Attaching Two Parts

Let's go over the various ways of attaching two parts. These consist of stud-and-tube—the classic LEGO method—as well as Technic pegs, cross-axles, and a variety of hinged parts.

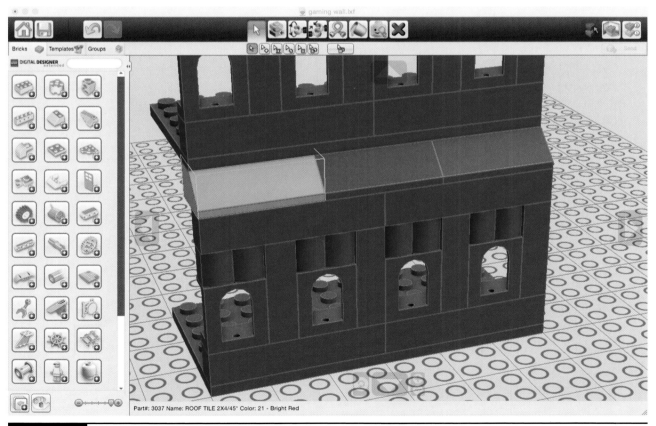

Figure 6-16 To change a brick's color, click on a brick with the Paint function selected.

Figure 6-17 First select the part you want to decorate; then click on Decoration to view a pop-up of options.

Stud-and-Tube

The classic LEGO building method involves attaching the top of one brick, which is marked with studs connecting to the bottom of a similar brick, which is hollow but for a series of friction-providing tubes. Simply drag one brick on top of another (Figure 6-18), and they will automatically snap together. It's the simplest connection system and one that we all very likely know about, if only from our own childhood.

Technic Pegs

Another common connection method involves using small pegs to connect two parts. It's a stronger bond than stud-and-tube, and therefore, it's mostly used in LEGO's robotics sets, namely, the Technic beams, which find their way into

Mindstorms boxes, among other product lines. If you want to attach a peg to a Technic part, simply drag the peg onto the appropriate mounting hole, and if it's aligned correctly, it will automatically attach. Figure 6-19 shows a Technic peg being inserted.

Cross-Axle

Cross-axles offer another Technic-friendly connection method. They come in varying lengths and offer more rotational friction than ordinary pegs. This allows you to use them either to immobilize a part that otherwise would rotate or to turn a wheel with a motor or power a gear assembly.

Because of their cross-shaped profiles, sometimes these parts have to be rotated first

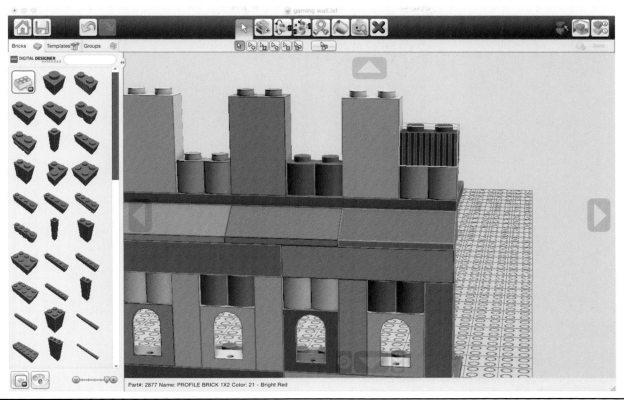

Figure 6-18   The classic connection system is called *stud-and-tube*.

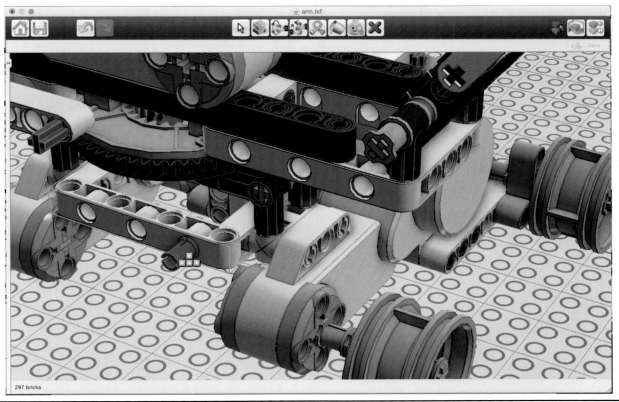

Figure 6-19   Inserting a Technic peg into an appropriate hole.

Figure 6-20 Cross-axles either make things move or keep them from moving.

in order to match the cross-shaped holes of the receiving brick. Figure 6-20 shows an axle being inserted.

Hinges

A number of parts have some sort of hinge— LEGO has a few different combos it uses. Simply drag one-half of a hinge to the other (such as the hinged dome in Figure 6-21), and they will attach automatically. In order to bend the hinge, you will have to employ one of two basic methods. Either you can bend it manually, simply by changing the angle of one of the elements, or you can use the Hinge Tool (detailed later in this chapter) to automate the process.

Ball-and-Socket

A less common method of attaching two parts, the ball-and-socket, mostly is seen in LEGO's Bionicle and Hero Factory lines. As one would expect, the method works much like the other

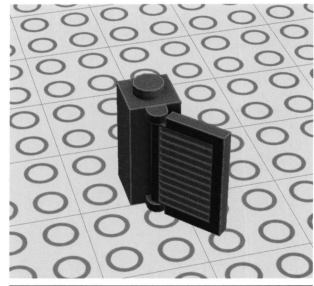

Figure 6-21 Attaching a hinge to its mate.

methods: you simply drag one part over to the other, and it pops right in, as seen in Figure 6-22.

Other

I know I'm missing a couple of other methods, for instance, putting a car tire on a rim. I

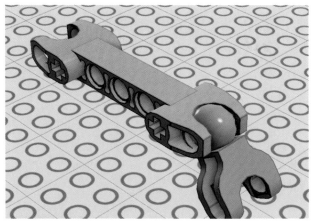

Figure 6-22 The ball-and-socket offers another way to attach two parts in LDD.

whether by using a group, a template, or through multiple selection—more on these techniques in the next section of this chapter. Once you have the parts you want selected, drag them on top of the target assembly. As you can see from Figure 6-23, the assembly to be added appears washed out when it's behind another element, simply to help you guide the two assemblies. When mounting connections such as pegs or studs, the contact point will get a green box around it, letting you know a possible connection has been made, When you're done, let go of the mouse pointer, and the two sets of elements are connected.

think you get the idea, however! Most of these techniques work much the same way.

Attaching Multiple Parts

Attaching two sets of parts works exactly the same way. Simply select the parts to attach,

The Physics of LDD

The LEGO Group added some rudimentary mechanical physics to LDD. If you attempt to rotate a cross-axle, for instance, it won't turn unless everything it's connected to turns as well (Figure 6-24). On real computer-aided design

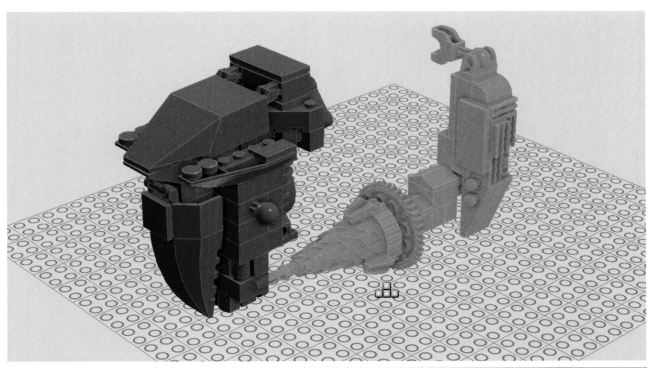

Figure 6-23 Select multiple parts to insert them simultaneously.

Figure 6-24   If you turn a cross-axle, everything connected to it turns as well.

(CAD) programs, engineers can make an entire mechanism work on screen. LDD, while really cool, simply isn't a professional-grade program costing thousands of dollars. LDD's physics, therefore, are cool but not always useful. Even assemblies that ought to move might not because the program thinks there's too much torque on the axle.

However, gravity doesn't exist in LDD, and a freely rotating element that ought to swing down will not. This presents both positives and negatives. On the positive side, you can make cool models that would never work in the real world. By contrast, if you ever tried to *build* one of these models in the real world, it would collapse. Good and bad, these glimmers of physics make me want more.

Interacting with the Grid

The final topic of this section is that lovely gray grid that comprises the workspace. It has a

bunch of circles on it as well as gridlines. As you might suspect, the grid conforms to standard LEGO dimensions: each square with four circles in it matches a 2 × 2 brick.

- Every element of the model must remain on the grid, and if you try to move the model off the grid, LDD will simply make the grid bigger.

- No part may be placed *below* the grid. If you try to move one there, LDD simply lowers the grid. If you have other parts sitting on the grid, they will find themselves raised up because of that one element at the bottom of the stack. If you take away that bottom brick, the grid pops back up. You can see this feature in action in Figure 6-25.

- The grid has no maximum size, but it does have a minimum size: the grid you see when you start the program represents the smallest it ever gets.

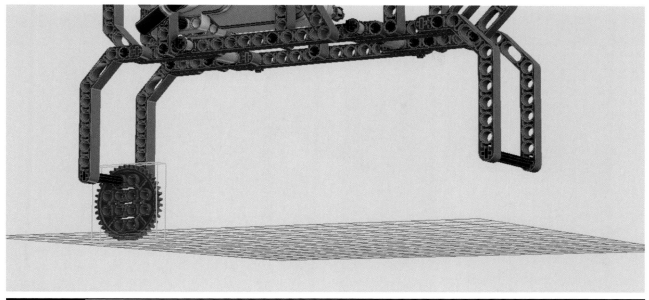

Figure 6-25 Nothing can be placed below the grid. If you try, LDD simply lowers the grid.

Advanced Building Techniques

Templates and groups are two ways to manage your bricks, and the tabs for these are located alongside the tab holding the brick palettes.

Templates

Templates store brick constructs for later, which can be dropped into models whenever needed. Say that you design a LEGO starship and you need a crew of nearly identical minifigs to populate it. Simply store a generic crewman as a template, and then pull in a few dozen of them.

Your designs are available in your LDD "Templates" pane no matter what file you're working in—*within that theme*. So regular LDD has one set of templates, LDD Mindstorms has another, and LDD Extended has a third, and when you're working in one theme, you won't be able to access the templates in the other themes.

When you click on the "Templates" tab, you are faced with a sidebar that shows the thumbnails of each design (Figure 6-26). This is what you can do:

- **Create a template.** To add something to a template, select the bricks you want, and click on the button on the bottom of the "Templates" tab and make the selection into a template. You'll see a thumbnail showing a miniature version of the items therein.

- **Access a template.** Click on the "Templates" tab to see your palette of templates. Simply click on the one you want, and it will appear selected in the workspace.

- **Delete a template.** There is a red X on each thumbnail. Click on the X for the file you want to delete. You will be prompted to confirm; then it's gone for good! Even if you quit without saving, that creation is also gone.

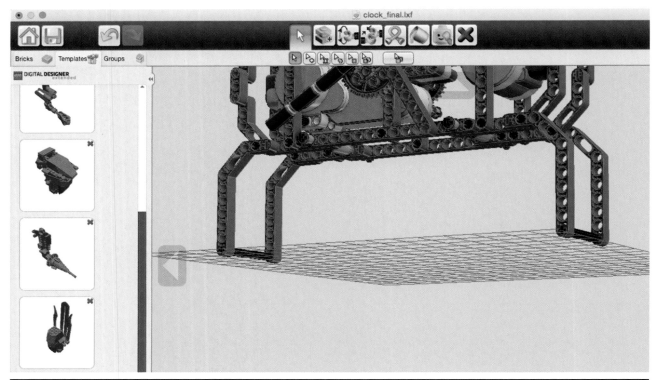

Figure 6-26 Templates store designs for future use.

Groups

Groups (Figure 6-27) are just that, groupings of bricks within that LDD model. They are primarily useful when you start making *really big* models, which begin to present certain challenges, particularly with regard to selecting specific elements that might be hidden back in the depths of the model and would be hard to get to. Here's an overview.

Group Features

You may find groups useful for the following reasons:

- Click on the group thumbnail to select all the bricks in the group, even if they have been moved around.

- If one or more bricks in the group get deleted, they will vanish from the group.

- Groups are not visible from other files, but if you import a file that contains a group, the group travels with the file.

- Groups can consist of layer after layer of subgroups, allowing you to control the bricks of a large creation by nesting several different layers of a group.

- You can easily split a group into two, as well as adjust the hierarchy of groups and subgroups to suit your needs.

Bending and Flexing

While not something you're likely to use all the time, bending and flexing are important techniques to have. They allow you to use parts that are more complicated than a mere brick. Hinge bending is basic. How do you move a door on its hinge without separating the door from its frame? Similarly, the Hinge Align tool

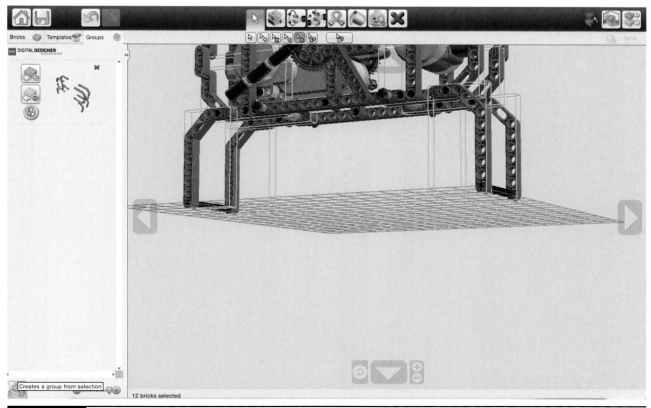

Figure 6-27   A group and attendant subgroups allow only the right bricks to be selected.

allows you to rotate a part so it lines up precisely with another part, allowing you to readily connect the two elements.

Hinge Tool

LDD offers countless varieties of hinges ranging from door jams to lidded boxes. These typically consist of two parts that click together in the real world, like the door and door frame. In LDD, these mostly work the same way.

In order for the hinge tool to work, an element has to be selected that can turn or rotate. Clicking on the Hinge tool results in a pop-up display to help you choose the direction of turn and the angle.

Figure 6-28 shows the dialog, which includes a couple of green arrows showing possible turns, with a blue wheel in the upper left-hand corner,

allowing you to manually turn the element—you don't pull on the element itself. You can also find a numeric angle indicator at the top of the screen, right below the "Select" button. You can simply type in the angle.

Hinge Align Tool

Sometimes you need to rotate an element to a specific angle, and you don't know the angle. You could tweak it a million times or simply use the Hinge Align tool, which allows you to align two mounting points.

Figure 6-29 shows an example. First, you click on one hole and then another that the element in question could actually reach when it rotates. The part will attempt to rotate and align the two items. However, if it's a complicated model and there is an obstruction that would reasonably

Figure 6-28   Click on an element that can be rotated, and hit the Hinge tool to move it.

Figure 6-29   The Hinge Align tool aligns two holes together.

prevent the rotation if it were a real model, the part will attempt to rotate and then return to its original location.

Flex Tool

Certain elements can be made to flex. The number is actually quite small and includes rubber "flex axles," hoses, and just a few other items. Figure 6-30 shows a tube that has been flexed.

You begin by attaching one end of the flex element to a mounting point—the element won't flex without one end attached to the model.

Then click on Flex tool and begin to bend the element around. Control isn't great, and sometimes it takes a few minutes to get the flex element attached to the second mounting point.

You're not likely to use this tool very often because only a small number of parts flex, but it's a great tool to have as an option.

Summary

Thus ends your LDD boot camp. In Chapter 7, I'll give the LDraw platform a similar look.

Figure 6-30   A tube that has been flexed.

CHAPTER 7

Building with LDraw

WHILE LEGO DIGITAL DESIGNER (LDD) sets the pace, LDraw does a lot of things right as well. Its gigantic library of fan re-creations of out-of-print LEGO bricks puts the competition to shame. Editors and other software solutions continue to evolve, with new features steadily improving the products. In this chapter, I'll share the ins and outs of both Bricksmith (Figure 7-1) and LDCad, the two most popular LDraw editors.

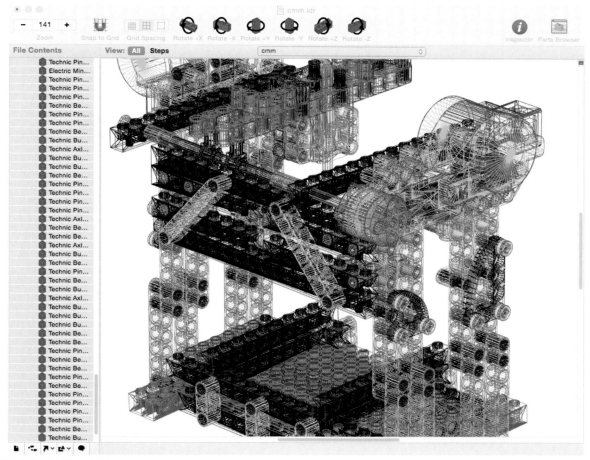

Figure 7-1   Bricksmith gives Mac users the full LDraw experience.

LDraw Editor General Features

LDraw editors share a number of common features, not only computer-aided design (CAD) standards but also shared conventions with other LDraw projects. In this section I'll go over a number of features that both LDCad and Bricksmith share.

LDraw Library

Of course, the most obvious commonality is their shared library. The LDraw library takes the form of the parts browser in Bricksmith (Figure 7-2), but every LDraw user sees the same parts, at least in a perfect world.

Some users prefer seeing a thumbnail of the element because they can spot it faster than in a primarily text-based interface. This sort of user will prefer the LDCad parts bin, which assumes that users will first browse visually before locking down the specific brick.

Figure 7-2   Bricksmith's parts browser lets you choose among LDraw's thousands of elements.

Color Selector

Unlike LDD, which attempts to keep the colors accurate for two of its three themes, LDraw editors let you choose any color for any brick. While Bricksmith presents the official colors as paint swatches, LDCad has a pleasing color wheel (Figure 7-3) that serves the same purpose.

Multiple Views

In the early days of LDraw editors, they would rotate around a model the way you can in LDD and Mecabricks. As a result of this limitation, it was easier to offer multiple views of a project so that you could see it from as many angles as possible. I like the views, and it's great to see different angles of the model as it comes together. Figure 7-4 shows a model open in Bricksmith with a primary pane in the center and three alternative views on the right.

However, LDCad and Bricksmith both allow that kind of mouse-based scrolling that offers a compromise between LDD's imprecise but forgiving three-dimensional (3D) navigation where precise angles aren't needed and old-school LDraw editors reliant on accurate measurements.

Precise Placement

In LDD, you really can't type in a number to position an element precisely in XYZ space. You must eyeball it, relying on LDD's snapping action that helps compatible parts snap together like two magnets attracted to each other. LDCad has implemented rudimentary part snapping, but it's far from the LDD experience. It boils down to a difference in audience. LDD aims for kids, who (apparently) can't be troubled to precisely align parts the way the adult LDraw fans often must.

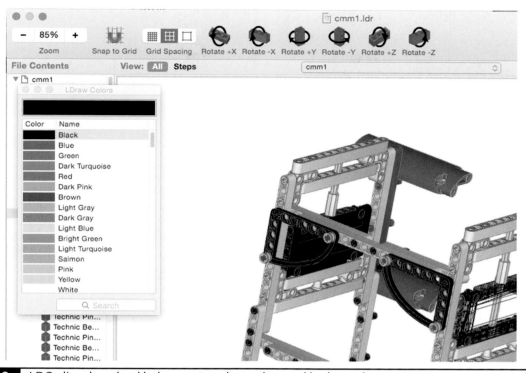

Figure 7-3 | LDCad's color wheel helps you to select colors and looks cool.

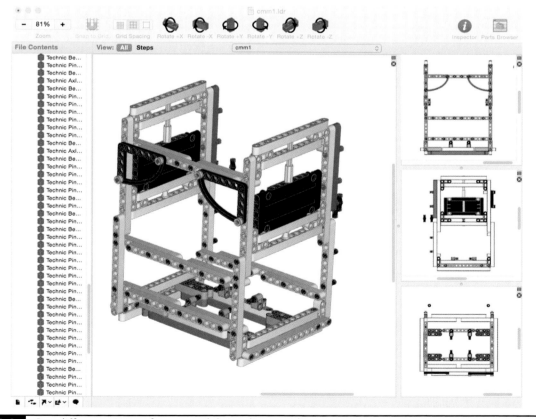

Figure 7-4 | See different views of your model simultaneously.

In the meantime, LDraw has something arguably better, and that is precision placement of bricks via numeric coordinates. Figure 7-5 shows the Part Inspector dialog for the currently selected element. Not only can you place the element via XYZ coordinates, but you can also use this dialog to rotate the part. Even crazier, you can also *scale* the part so that it's bigger or smaller relative to the rest of the model.

Submodels and Steps

A very clever feature built into Bricksmith and LDCad, Steps, breaks a project down into a series of smaller submodels, each with its own steps built in (Figure 7-6). For instance, if you're making a humanoid robot, the two arms and two legs might be submodels unto themselves.

Building in these divisions allows you to create a parts list for each step, a useful strategy when dealing with really big models. It also gives you the focus to work on a vital subcomponent of a model. For instance, say you need to make a complicated gearbox for a model. Being able to keep the gearbox separate, as its own file, ensures that nothing gets messed up.

Perhaps more important, dividing a model into a series of steps and substeps induces the builder to think in terms of building instructions. I cover these vital how-tos in Chapter 9. You can even export every step as its own .LDR file, a feature I wish LDD had as well.

Complexity

Compared to both LDD and Mecabricks, you will find LDraw to be extremely complicated.

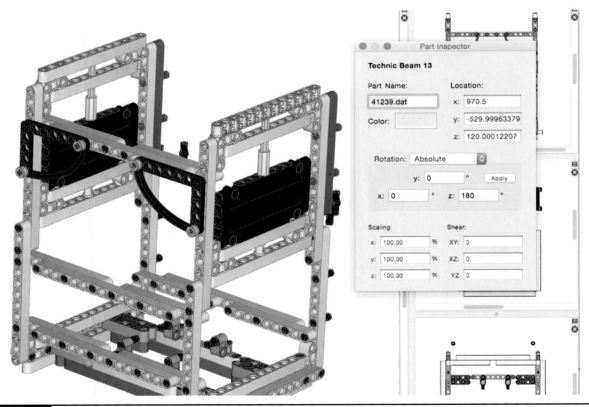

Figure 7-5   Type in coordinates to change a part's location.

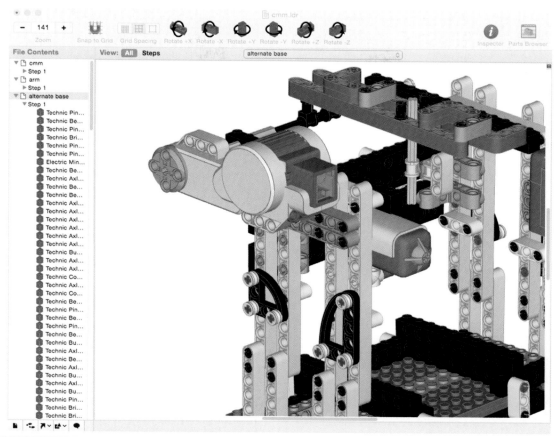

Figure 7-6 Smart builders break their projects down into steps.

Part of it is the fact that LDraw is mostly focused on adult fans who have the patience to learn a complicated system but who also demand a more sophisticated application than a child might. LDD was designed for children to use, and Mecabricks, as a website, must offer a simpler interface than a full-fledged application. LDraw apps were designed by hobbyists for hobbyists.

This is not to say that most people couldn't use Bricksmith and LDCad. You'll encounter a steeper learning curve than the other two apps, but by reading the help files and forum entries, as well as watching instructional videos, you'll soon be making models.

Building with Bricksmith

Bricksmith pretty much owns the category of Macintosh LDraw editors. In the following paragraphs I'll cover the interface and menus.

The Top Menus

The top menu of the program assists you with the following tasks.

File

This includes all the usual features of File menus, such as Open, Save, Save As, and so on. The most LEGO-specific option is to Export Steps, which conveniently outputs every step of the project as its own .LDR file.

Edit

Cut, Paste, Duplicate, and other tools work as you would expect. You also have the option to split a step into two.

Tools

This menu holds a couple of useful widgets. Dimension gives you the actual dimensions of the model in different measurements—centimeters, studs, and even Legonian imperial feet, the scale if minifigures were human height. Figure 7-7 shows the dimensions of a Bricksmith model. You can also reference a parts list, a feature missing from LDD. Finally, you can change the grid setting, a duplicate of the button-bar option.

View

View changes zoom levels and that sort of thing. It also toggles you between viewing the LDraw model as one giant step and as a series of smaller steps.

Piece

The Piece menu is an oddity and just shows the various brick color options.

Model

The Model menu has a ton of useful tools. You can add new models or new steps, insert parts into the current model, or draw shapes, leave comments, and insert flex elements. My favorite tool, however, is a minifigure generator (Figure 7-8) that greatly simplifies adding those cheerful figures to your project.

The Workspace

In addition to the top menu, you'll also need to access options located in the workspace area. These include Mouse tools (callouts A–G in Figure 7-9) as well as the normal option buttons (callouts H–K in the figure). Also see the list of shortcuts in Table 7-1.

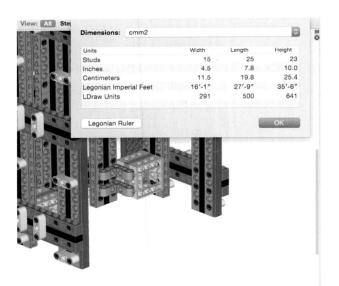

Figure 7-7   Measure your model in Legonian feet.

Figure 7-8   Generate your own minifig using Bricksmith's figure creator.

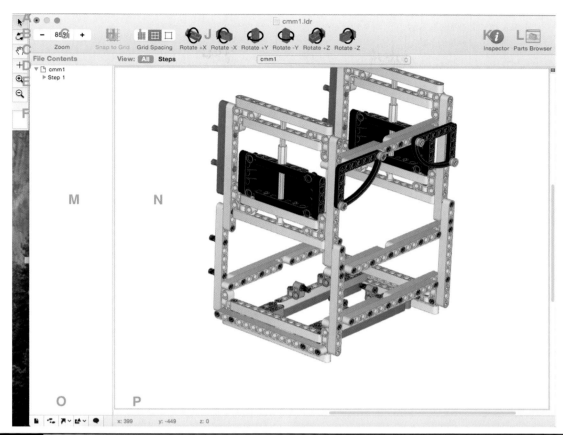

Figure 7-9 Bricksmith's menus contain a lot of options to help you with your build.

A. Selection

This works pretty much the same as the other platforms. You click on an element, and it becomes selected. Rotate Model puts the model in a mode where it can be rotated directed by your computer mouse. Pan Model works the same way.

B. Rotation

Rotate the model freely with the help of your mouse.

C. Panning

This tool helps you to move from side to side.

D. Center

This tool centers the model on the screen.

E. Zoom In and Out

These buttons zoom the camera in and out, as one would expect.

F. Part Color

This swatch reminds you of your current brick color.

G. Zoom

Control the zoom amount with a slider and numeric field.

Table 7-1 Bricksmith Shortcuts	
Click+Drag or Arrow Keys	Move element
SHIFT+Click+Drag	Constrained movement
OPTION+Click+Drag	Make a copy and move it
OPTION+Arrow Keys	Move element on third axis
X, Y, or Z	Move on that axis
SHIFT+X, Y, or Z	Negative movement on that axis
Click	Select an element
SHIFT+Click	Multiple selection
CMD+Drag	Change camera angle
SPACEBAR	Pan
CMD+OPTION+Click	Center on clicked point
CMD+OPTION+Drag or CMD-Scroll	Smooth zoom in and out
CMD+SPACEBAR	Zoom in
OPTION+SPACEBAR	Zoom out
0	Three-quarters view
2	Bottom view
4	Left view
5	Front view
7 or 9	Back view
8	Top view

H. Snap to Grid

This helps to keep the layout organized by automatically snapping each part to increments of an invisible grid.

I. Grid Spacing

This tool changes the coarseness of the grid.

J. Rotate Model

These six buttons control the camera's movement along XYZ axes in 45-degree increments. You must have a brick or bricks selected to make them move with this tool.

K. Parts Inspector

The Parts Inspector gives you details about the selected brick. You can see an example in Figure 7-9.

L. Parts Browser

Your faithful library of LEGO elements—I already covered this in Chapter 5.

M. File Contents

This pane helps you to keep track of your steps. You can also see comments embedded in the steps (if any), so if you're collaborating with someone else, you can share thoughts that way.

N. Build Area

Work on the model in this pane.

O. Annotation Toolbar

This small toolbar allows you to comment and add shapes such as circles and squares to the model.

P. Mouse Coordinates

This indicates your mouse pointer's location in XYZ space.

Building with LDCad

LDCad offers considerably more complexity than either Bricksmith or LDD. Its creator, Roland Melkert, continues to add features to the platform. However, he also produces great help files and videos, greatly aiding in learning the application's ins and outs.

Once you have the software and part library installed, it's time to get to building. The following tools are featured on the LDCad

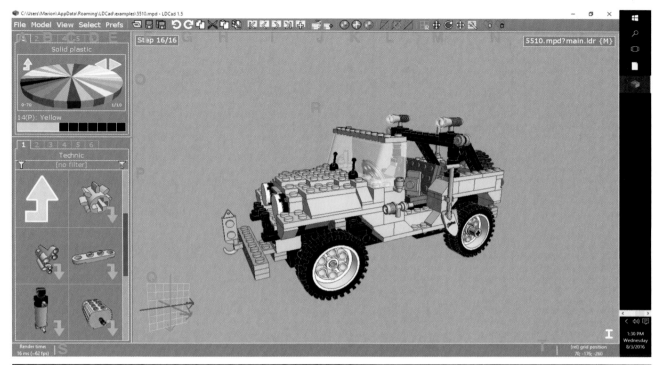

Figure 7-10   LDCad's interface packs in a lot of tools.

interface. Follow along with Figure 7-10 as I go over them and see Table 7-2.

A. File

The typical file system with different ways to save your work.

B. Model

This menu gives you many options for managing the models that make up this LDraw file. You can designate main models and submodels, as well as giving you options for adding annotations such as shapes and text to the design. This menu also governs adding sub-assemblies created by other programs, such as flex elements created by LFlex.

C. View

This opens up the parts bin and color wheel windows as needed.

D. Select

LDCad offers an impressive assortment of selection tools, including selecting similar part numbers and colors, as well as selections in multiple steps.

E. Prefs

Adjust the preferences to make LDCad work better for you.

F. File System

The file system works as one might expect, and allows you to open LDraw files as well as to save all files.

G. Undo and Redo

This works as one would expect.

Table 7-2 Shortcuts	
CTRL+SHIFT+S	Save all changed LDraw files
F11	Show/hide snap info points
UP/DOWN or mouse scroll wheel	Navigate menus
ESCAPE	Return to the parent group
HOME/END	Go to the beginning or ending of the group
LEFT/RIGHT or mouse scroll wheel	Rotate the color wheel
CTRL+mouse wheel	Tilt the color wheel
CTRL+left mouse button	Apply the color to the selection
1–6	Switch to Parts Bin numbers 1 through 6
F	Add the current element or color to your favorites
SHIFT+F	Unfavorite
F 3	Open the Parts Bin filter dialog
INS	Insert the part the mouse pointer is on
Double-click left mouse button	Insert the part into the model at 0, 0, 0 or replace a selected element
Single-click left mouse button	Select item
CTRL+left mouse button	Deselect item
SHIFT+left mouse button	Set element as the current working part
CTRL+D	Duplicate selected part
CTRL+INS	Append a building set
CTRL+HOME	Reset the selected element XYZ to match the grid orientation

H. Copy and Paste

Industry standard all around.

I. Navigate Steps

As with Bricksmith, users are encouraged to build models as a series of subassemblies called "steps." These buttons guide you back and forth through the steps.

J. Model

Go to the main model of the current LDCad file or add a new model.

K. Grouping

Click to add or remove bricks to or from a group.

L. Hiding

Hide the selected parts, undo the most recent hide, or unhide all.

M. Numerical Movement and Rotation

As with Bricksmith, LDCad users need to be able to have precise control over their elements.

N. Favoriting

Save favorite bricks and colors with these buttons.

O. Color Wheel

Select your color visually by clicking on the wheel. You can choose among several different color palettes.

P. Part Library

The library's part selection offers nested menus with all the LDraw parts one could ever want—assuming they've been installed!

Q. XYZ Indicator

Know what angle your model is pointing by looking at this indicator.

R. Model Build Window

This is where the magic happens! LDCad's tiny menus (in my opinion) save a lot of real estate.

S. Render Time

This amount of time presumably increases as you build gigantic models. Right now it's pretty much instant.

T. Mouse Position

This is an important piece of infomation when you are using a coordinates-based system.

Scripting and Animation in LDCad

In terms of sophistication, LDCad stands out from the other applications I've covered in this book. Case in point: scripting and animation. LDCad has the ability to perform simple animations to illustrate how a model works. It has videos of meshing gears, made possible by incrementally rotating the axles to simulate how they might work together on a real model. This technology puts LDCad head and shoulders over the competition, at least in this regard.

On the downside, the scripting might offer a considerable challenge to some. For instance, you must familiarize yourself with the syntax of Lua scripting (lua.org) before you can do anything. It's about as difficult as the LDraw file syntax (described in Chapter 10), and if you can wrap your head around one, you certainly can learn the other. I encourage you to give it a try even if it's outside your normal area of interest. If you're interested in scripting, study the script API that LDCad designer Melkert has on his site (www.melkert.net/LDCad/docs/scriptAPI).

Summary

In Chapter 6 you learned about the LDD platform, and in this chapter, you discovered what makes LDraw unique. In Chapter 8, I'll walk you through the web app that is shaking up the virtual LEGO building world!

Building with Mecabricks

IN THIS CHAPTER, you'll get a sense of the Mecabricks interface and the program's capabilities. As a web app, it will provide you with a much different set of advantages and disadvantages compared with the traditional downloaded applications.

On the one hand, you can use Mecabricks from anywhere you might be in the world, accessing your designs and resuming your projects even if you're a thousand miles away from your usual desk. On the other hand,

this might also be a problem because Internet connectivity is no more perfect than any other technology. It has all the usual disadvantages of not storing one's own data. What happens if the Mecabricks server coughs up a hairball and you lose your project? What if you lose your login info or the service shuts down? These are challenges that LEGO Digital Designer (LDD) and LDraw users don't have to deal with. On the upside, Mecabricks has a wonderfully simple interface (Figure 8-1) and offers a very robust building experience.

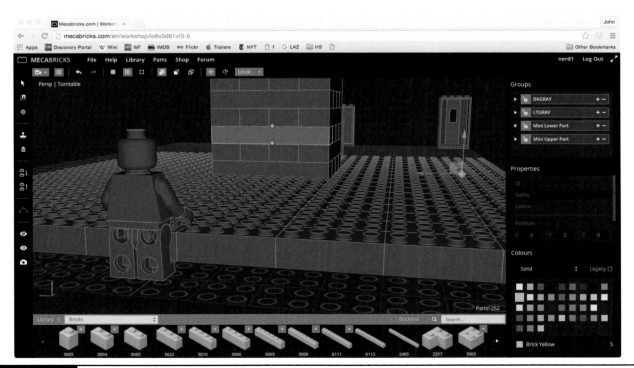

Figure 8-1   A model takes shape in the Workshop.

In the following sections I'll guide you through the interface and its features, beginning with the program's home page and moving on to the four main-menu items: the Workshop, which serves as the actual building program; the Library, where users share their latest creations; Parts, which helps users to understand all the elements available; and the Forum, where questions are answered and problems addressed.

Navigating Mecabricks

When visiting the website Mecabricks.com, you are faced with the following interface.

The Home Page

Like all classic home pages, Mecabricks.com's main page describes the product for those who are visiting for the first time. On the top part of the home page there is a menu bar and a beauty shot of a digitally built model on someone's phone with a physical model next to it. There's a call to action that serves the same function as the "Workshop" button, and all the call to action does is bring the user into the Workshop. Finally, there is a small ad at the lower right-hand corner. Once you scroll down past the top page, you get a display of models from the Library as well as feeds from the @Mecabricks Twitter account and an invitation to "Like" their Facebook page.

Figure 8-2 breaks down the home page. Follow along with the caption to learn about each item.

A. Home Page Link

As usual, the site logo may be clicked on to return one to the homepage.

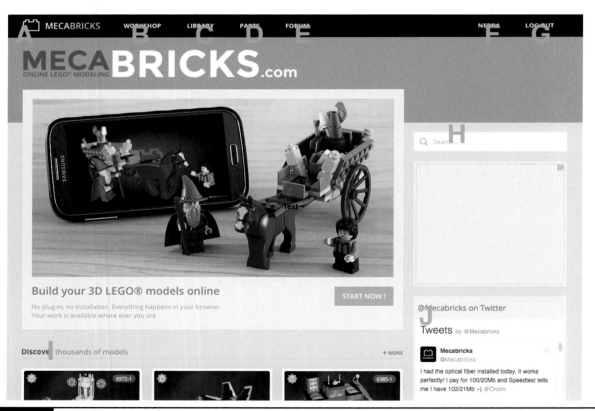

Figure 8-2 Mecabricks' interface is easy to interact with.

B. Workshop

This is the most important option because the Workshop is where you build models. I'll get into the specifics of this interface later in this chapter.

C. Library

The Library holds all the models Mecabricks has built, some of which are public and some private, meaning that other people can or can't see what's being worked on. However, other parts of the site refer to the Library to mean the Parts menu.

D. Parts

Parts manages the vast array of LEGO elements that may be built with the program.

E. Forum

Mecabricks' Forum is solid, with bug reports, building tips, and a bunch more information available.

F. Screen Name

The user's screen name, if he or she is logged in.

G. Login or Logout

Login or logout, depending on the user's status.

H. Search

Search models in the Library.

I. Library Examples

Library examples, with a bunch more visible as you scroll down.

The Workshop

Next, let's take a look at the main area of Mecabricks—the Workshop (Figure 8-3). It appears rather complicated, with numerous cryptic icons, so let's go through each item.

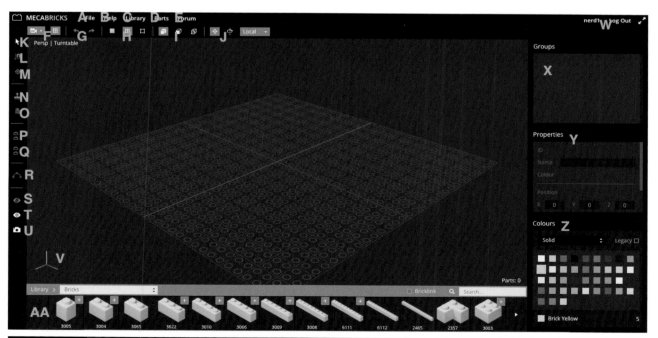

Figure 8-3   Mecabricks' interface offers numerous options for manipulating bricks.

A. File

This works as one might expect and allows you to load and save Mecabricks builds.

B. Help

While not a fully fledged help system, Mecabricks does offer a few important tips for newbies.

C, D. Library, Parts, and Forum

These work as per the menu bar on the home page.

F. Camera Control

This rotates the camera (the apparent viewpoint of the model) around the model. You can also choose between orthographic projection and normal perspective.

G. Undo and Redo

These operate as one would expect.

H. Grid Selector

These three selectors specify the resolution of the Workshop grid.

I. Rendering Selector

Select from three styles of rendering. There is default normal rendering style, "shading with edges," which takes up a smaller amount of memory while sacrificing some realism of the bricks. Finally, for systems that are *really* maxed out, there is "wireframe rendering," which takes out all the polygons and leaves in the edges.

J. Translation and Rotation

"Translation" in geometry means moving without rotating. Well, rotating is obvious,

right? "Global" means that the brick's XYZ coordinates remain the same, but the workspace rotates. "Local" means that the brick rotates, but the workspace does not.

K. Selection Tool

Choosing this tool gives you a pop-up window of options such as "multiple selection" and "area selection." It's pretty much the same set of selection tools as LDD.

L. Snap Tool

This tool allows you to connect two parts that would attach together in the real world, for instance, a pair of LEGO bricks. I'll show you how to do this later in this chapter.

M. Pivot Point

This tool helps to set the axis on which the brick or bricks will rotate.

N. Clone Tool

This clones the selected part, placing it on top of the original. Note that this is impossible to do in LDD, which prevents two bricks from occupying the same space.

O. Delete

This deletes the currently selected brick or bricks. You can also hit the DELETE key. Either way, you'll be prompted to confirm the deletion.

P. Make a Group

Mecabricks has a smiliar grouping system as LDD, where bricks can be added to logical or arbitrary groups. When the group is clicked on, all the bricks of the group are selected. When you click on the lock, the group's individual bricks cannot be modified separately, though it can still be moved.

Q. Ungroup

A selected group will ungroup when this option is clicked.

R. Flex Tool

This tool sets Bézier curves for flexible elements, allowing a flexible element such as a hose or flex axle to curve between two points.

S. Hide Part

The selected becomes invisible and cannot be modified.

T. Show All Parts

All hidden elements return to their original positions.

U. Screen Capture

This takes a snapshot of the Workshop grid at standard screen resolution. Any part of the model not visible in the pane will not be visible in the screen shot.

V. XYZ Indicator

This extremely handy pointer tells you which direction is which. In the LEGO world, the Y-axis is up and down, the X-axis is left and right, and the Z-axis is toward you and away from you.

W. Login Status

This indicator either urges you to log in if you haven't already or displays your screen name if you are logged in.

X. Groups

This pane lists all groups found in a particular model.

Y. Brick Properties

This pane details the part number and color of the currently selected brick.

Z. Colours

This color selector, similar to LDD and LDraw, allows the user to choose among all LEGO standard colors.

AA. Parts Library

The bottom of your screen consists of a pane with the parts palette visible. It defaults to the standard LEGO bricks palette but can be switched to any of Mecabricks' palettes.

While I know you're excited to get to building, I still have to finish describing the interface. Next up, the Library!

The Library

Proof of Mecabricks' awesomeness may be found in the Library, where users store and share their models. It's not a complicated system, and it has a number of search mechanisms in place to help users find models. Figure 8-4 shows how you can select from the variety of search methods. You can either browse categories or sort what you see by a number of criteria. Follow along with the figure caption.

A. Category Selector

You can choose between all models, sets, minifigures, and renderings. "Sets" refers to models rebuilt from official LEGO products.

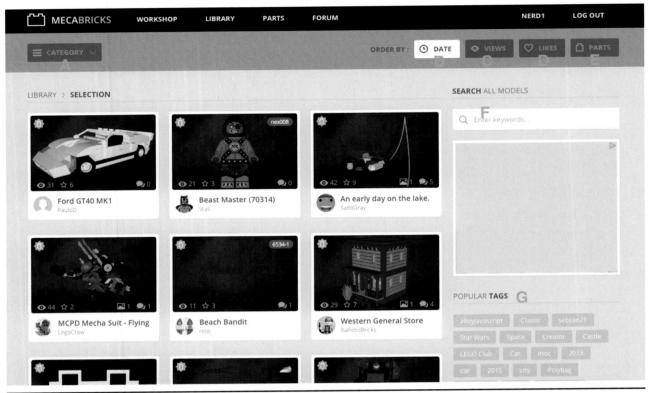

Figure 8-4   The Library holds every Mecabricks model, public and private.

B. Sort by Date

How recent is the model?

C. Sort by Views

Sort by number of views.

D. Sort by Likes

Sort by number of likes.

E. Sort by Elements

Sort by number of elements.

F. Keywords

Search keywords.

G. Browse

Browse by tag.

Once you've selected a model, you get three view options, as seen in Figure 8-5: 3D View, Renderings, and Editor.

- **3D View.** This resembles the model in the Workshop, and you can navigate around the model as you normally would. However, in 3D View, you cannot edit the model.

- **Renderings.** This view only populates if renderings have been made of the model, and not all models found in the Library have them. Renderings are a way of displaying virtual models, where all the polygons that make up the model are redrawn at a higher resolution, with simulated lighting and other effects that make it seem more realistic. In Chapter 10, I'll introduce you to some third-party rendering programs.

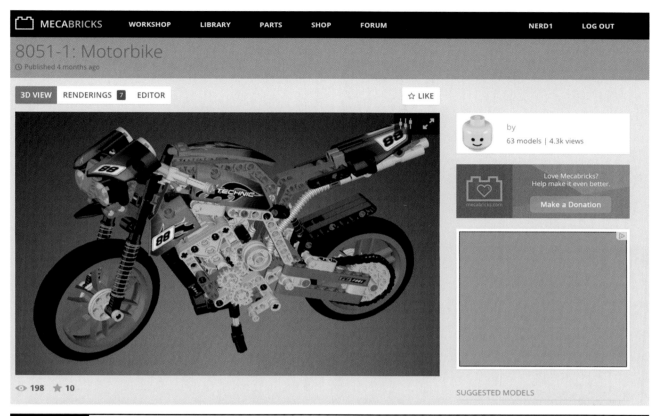

MECABRICKS   WORKSHOP   LIBRARY   PARTS   SHOP   FORUM   NERD1   LOG OUT

8051-1: Motorbike
Published 4 months ago

3D VIEW   RENDERINGS 7   EDITOR   ☆ LIKE

by
63 models | 4.3k views

Love Mecabricks?
Help make it even better.
mecabricks.com   Make a Donation

198 ★ 10   SUGGESTED MODELS

Figure 8-5   You can view a model in the Library one of three ways.

■ **Editor.** Clicking on this link brings the model into the Workshop, where it can be modified and saved to your Mecabricks account.

Below the model's image, you'll find more ways to learn more about the model. Figure 8-6 shows what you can find out.

■ **Views and likes.** These show the model's popularity, which gives you a sense of how good it is.

■ **About.** The default view, About consists of a statement by the builder as well as some tags that he or she selected.

■ **Inventory.** Ever wondered what parts make up a cool model? Wonder no more because the Inventory automatically tallies up the parts list of the model and displays it under

the "Inventory" tab, along with download links for comma-separated values (CSVs), a standard database format, as well as Bricklink-friendly XML. Bricklink is a cool site for buying and selling parts, so if you're looking to buy parts, this is a great way to upload your wish list to the site.

The next item, Share, invites you to share the model on your blog or social media site. You can either grab a link or an embed code. Icons to the right allow you to share the project via Twitter and Facebook.

At the bottom of the page are comments, if any.

Parts

You may be surprised by how much detail Mecabricks includes. For instance, it keeps track of whether each part is found in LDD's brick palettes. I covered Mecabricks's unique solution for organizing its parts (Figure 8-7) in Chapter 5, so you already know what's going on.

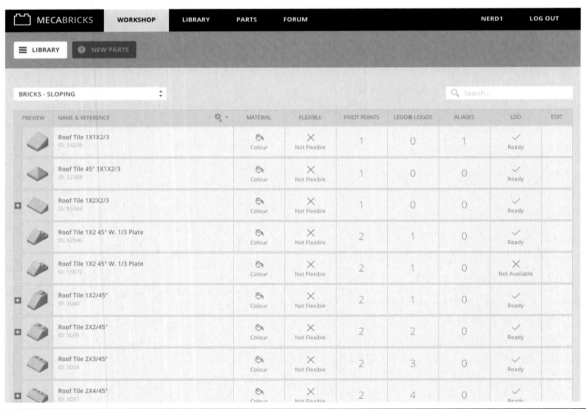

Figure 8-7 Mecabricks' parts menus feature every part in a table format.

Forum

Mecabricks features a very helpful Forum area (Figure 8-8) where creators answer users' questions about bugs, current or prospective features, and building techniques. It turns out that the best way to search for specific keywords is to use Google or some other search engine because Mecabricks' Forum lacks a build-in search engine.

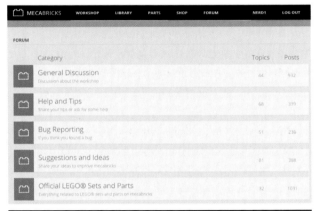

Figure 8-8 Mecabricks' Forum will answer your questions.

Building in Mecabricks

As you might expect, the basic mechanics of building with LEGO bricks are pretty much the same from system to system. That said, Mecabricks does have a couple of unusual advantages and limitations. Let's go over some basic techniques.

Set the Grid and Ground

The build surface, as with LDD, resembles a ghostly LEGO plate, allowing you to gauge the size of unbuilt models by counting the studs. You can make the "ground" (as Mecabricks calls the LEGO plate) appear and disappear (Figure 8-9). You can also change the "grid resolution,"

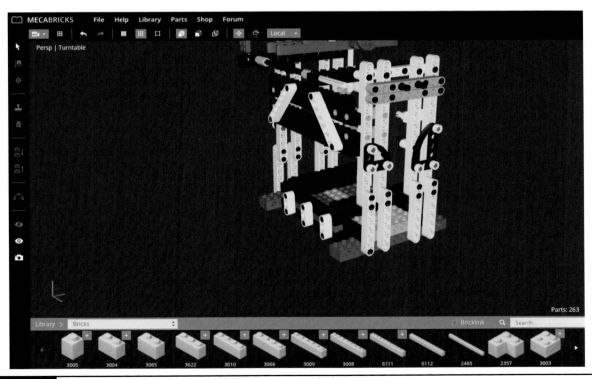

Figure 8-9 Make the ground appear or disappear as needed.

which govern how far a part moves with each click. The widest grid equals 2 LEGO studs, or 16 in Mecabricks' coordinate system. The middle level is 4 (half a brick's width), and the smallest grid is one-eighth of a brick's width.

Camera Control

Click on the workspace and drag to move the model around, changing all three axes (Figure 8-10) and zooming with the mouse's clickwheel. You have a choice between a "trackball-style" camera movement, which readily zooms around all sides of the model, or "turntable," which mostly rotates with the bottom staying down.

Mecabricks' key commands are similar to those found in LDD and LDraw, but they do not require hitting CTRL (Table 8-1).

Adding New Bricks

Let's begin a Mecabricks' project. Find a brick you like, click on it, and it appears, selected, on the ground plate at location 0, 0, or the exact center of the plate. Yank on the arrows sticking out of the brick (Figure 8-11) to move it around. Pretty simple!

One big difference between LDD and Mecabricks is that in Mecabricks, two elements can occupy the same space. I added another brick, and it appeared at 0, 0 as well. You can see the old brick and the new one in Figure 8-12.

Similarly, if you have a brick selected and add a new one, it arrives at the same coordinates as the one you have selected. If they happen to be the same style of brick, you might not even notice that there are two of them. If no brick is selected, the new element goes to 0, 0.

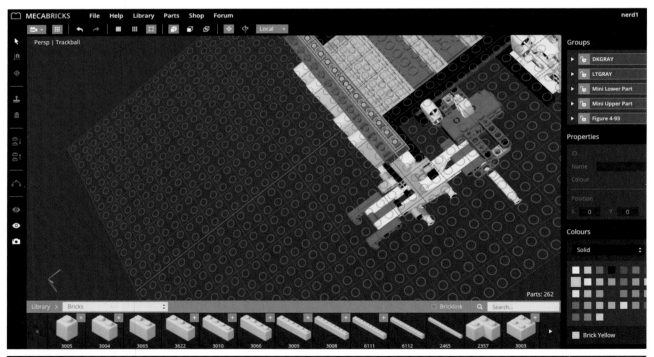

Figure 8-10 Move the camera around with your mouse.

Table 8-1   Camera Control Hot Keys

Command	Action
A	Select or deselect all
B	Select parts located in the area
C	Screenshot
D	Clone selection
F	Flexible Part tool (only active for flexible parts such as rubber bands, flex tubes, etc.)
H	Hide selection
ALT+H	Show everything
J	Group selection
ALT+J	Ungroup selection
M	Toggle between translation and rotation gizmo
P	Select pivot point
Q	Toggle among small, medium, and large transformation grid
R+(X or Y or Z)+numerical value	Rotation along selected axis in the global or local space
S	Select snap point
T+(X or Y or Z)+numerical value	Translation along selected axis in the global or local space
V	Toggle between global and local space
W	Display selection tools
Z	Toggle among shaded, shaded with edges, and wireframe mode
CTRL+Z	Undo
CTRL+Y	Redo

Table 8-1   Camera Control Hot Keys *(continued)*

Command	Action
Numpad 8 or 2	Translate along Z-axis
Numpad 4 or 6	Translate along X-axis
PAGE UP or PAGE DOWN	Translate along Y-axis
← or →	Rotate selection 45 degrees along Y-axis
↑ or ↓	Rotate selection 45 degrees along X-axis
HOME or END	Rotate selection 45 degrees along Z-axis (keys next to PAGE UP and PAGE DOWN keys on an Apple keyboard)
ESC	Deactivate selected tool/cancel rotation or translation
CTRL+Click	Add or remove a part/group to the selection
Numpad 5	Toggle between perspective and orthographic projection
Numpad 9	Toggle between turntable and trackball orbit style
Numpad 0	Reset the position of the camera
CTRL+Numpad 3	Left view
Numpad 3	Right view
CTRL+Numpad 1	Back view
Numpad 1	Front view
CTRL+Numpad 7	Bottom view
Numpad 7	Top view
SPACEBAR	Change camera orientation
DEL or BACK	Delete selection

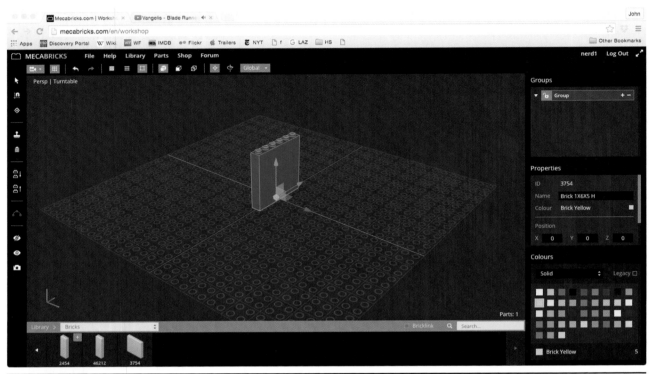

Figure 8-11 Click on a brick, and it appears on the Workshop's ground plate.

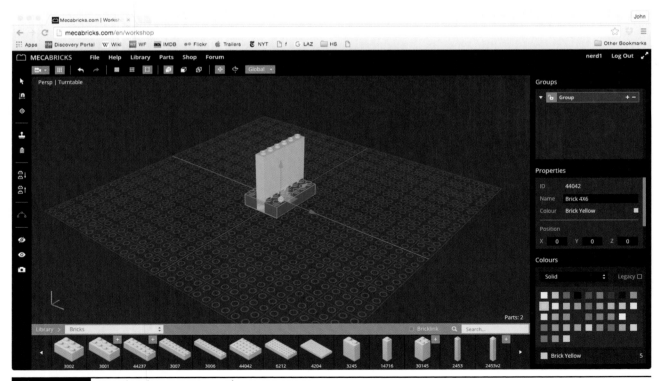

Figure 8-12 Two bricks can occupy the same space.

Cloning

You can duplicate an element or elements simply by selecting the part or parts and clicking on the "rubber stamp" icon. As usual, the new bricks appear behind the old ones, selected, and you can use the Translation tool to drag it to a new location. Note the red outlines on the double bricks, revealing that there is a duplicate behind it (Figure 8-13).

Rotation and Translation

Mecabricks offers two ways to move bricks, Rotation and Translation. Before I describe them, however, I want to talk about two other options, Global and Local. What this means is that when the part has been moved or rotated, is its XYZ positioning still the same as it was before, or has it changed for that element? Most of the time you'll want to stick with the default of Global, but Local is an option if you want to change the angle of movement.

Rotation

This rotates an element without moving it. Mecabricks' rotations snap to 45-degree increments, giving you a greater chance of getting elements to snap together. When you click on "Rotation," a series of three circles

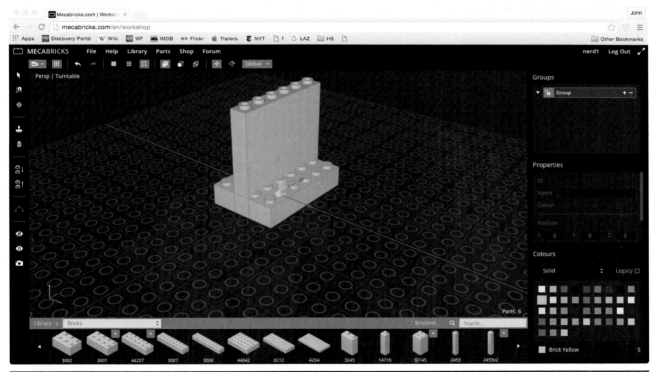

Figure 8-13   Clone a duplicate of the selected bricks.

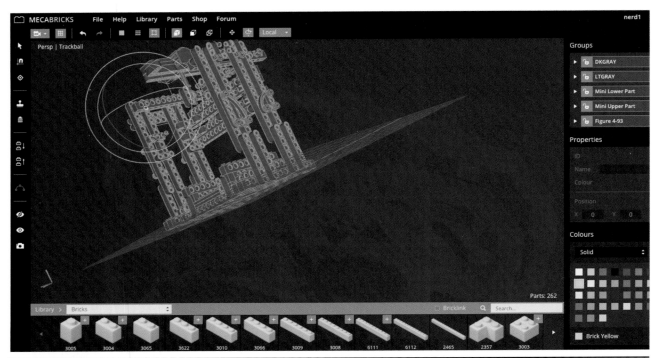

Figure 8-14 Rotate rotates without moving.

appears around the element (Figure 8-14) showing the axes along which it can be rotated. Simply click on one of the circles and move it to rotate along that axis.

Translation

This moves elements without rotating them. Figure 8-15 shows a move in process. You can see that three colored arrows project from the

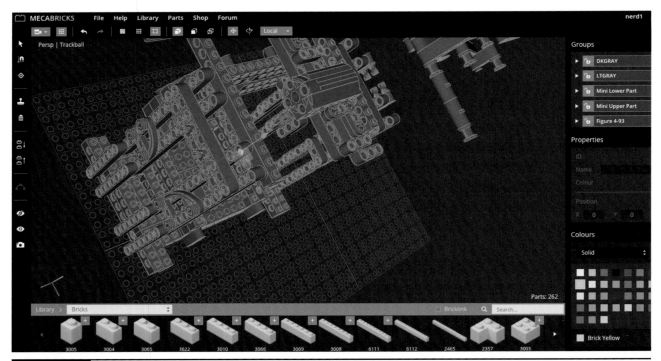

Figure 8-15 Translation moves without rotating.

element, corresponding with X, Y, and Z. To move the element, simply click on one of the arrows, and drag it in the direction you want.

You can also move a brick by first selecting it, and then entering the desired coordinates into the "Properties" pane. Elements move in fractions of a 2 × 2 brick's width depending on which Grid setting you chose.

Grouping

Like LDD's and LDraw's groups, Mecabricks makes submodels that can be selected as a group. Figure 8-16 shows three 2 × 2 bricks that I've collected into a group. You can add groups to other groups, and ungroup any groups as needed. To add a new brick to an existing group, select the brick to be added, and then hit the plus ("+") button on the group's listing.

Unsurprisingly, to remove one or more elements from a group, select that item or items and click on the minus ("–") button on the group's listing.

Hinges

Mecabricks lacks LDD's slick Hinge tool, but it does have a slightly more complicated tool that can potentially be much more powerful. You can set a pivot point on an element, and when you rotate the object, it turns on that pivot rather than the center. Simply select the brick you want to rotate, then click on the "Pivot Point" button, and select the part of the brick you want to serve as the axis (Figure 8-17).

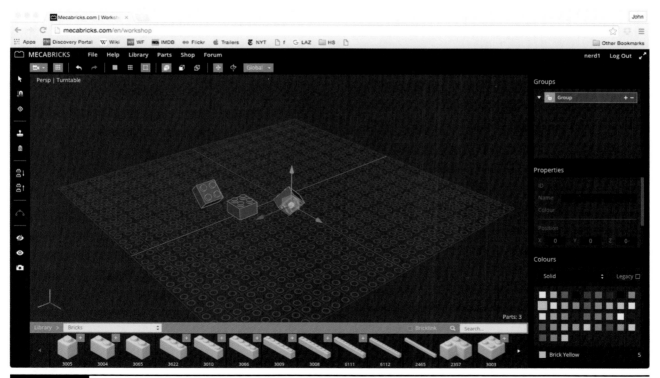

Figure 8-16 Groups gather together subsections of a model.

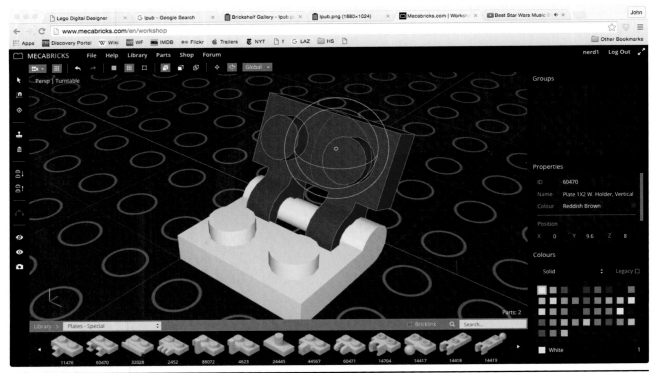

Figure 8-17   Hinges rotate with the help of a pivot point.

Flexing

Mecabricks' Flex tool uses a clever system to manage its flex elements. Add a flex element to the workspace, and then, while it's selected, click on the "flex" icon, and the element will turn into a Bézier curve (Figure 8-18).

You can modify the curve by double-clicking on one of the two toggles. When you're done, hit ESCAPE, and the tube reappears. Figure 8-19 shows the newly curved tube. If you don't get it quite right, just go back into the Flex tool and double-click on one of the toggles, the same as you did before.

Snap

The Snap tool is one of the coolest innovations Mecabricks has brought to the virtual LEGO building world. The idea is to connect two parts that would ordinarily connect, such as a door into its jam or two bricks stacked on top of each other.

When you select an element in Mecabricks, you'll notice a number of yellow dots, visible in Figure 8-20. These are snap points that help to guide the various bricks together. For instance, a classic 2 × 4 brick has two surfaces that can be snapped, the top and the bottom.

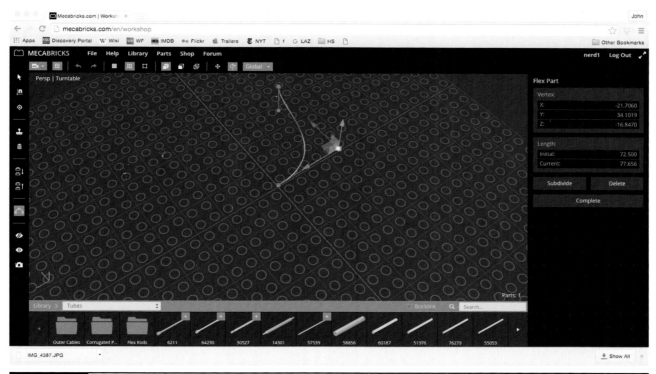

Figure 8-18 Modify the Bézier curve to change the tube's curve.

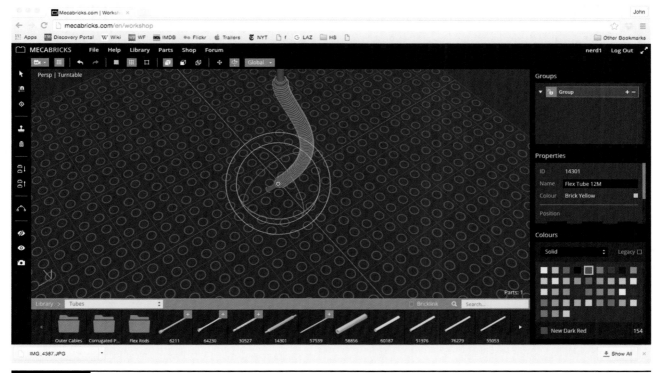

Figure 8-19 Flex elements can be adjusted to fit the needs of the project.

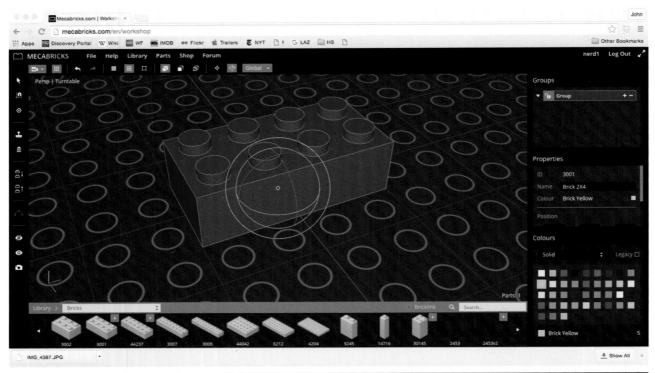

Figure 8-20 A classic 2 × 4 brick has two snap points, marked with yellow dots.

Interacting with Other Platforms

With all the talk about LDD, LDraw, and Mecabricks, one might ask how well the three systems work together. I'll tell you right off that Mecabricks does not interact with LDraw in any way. LDraw uses a proprietary measurement system for its polygons that is different from both LDD and Mecabricks. LDD and Mecabricks would have to adapt thousands of parts to translate these models, and they're focusing on different things.

Even between LDD and Mecabricks there are conflicts. Mecabricks' angles are right handed, whereas LDD angles are left handed. However, they are much closer together.

Saving to Mecabricks' Library

Don't forget to save! If you close the website before saving, all your progress will be lost. If you save to the Library, you'll have a number of options, such as selecting tags or deciding whether to share your model publicly. Figure 8-21 shows the Library page.

Figure 8-21 Save frequently!

Importing

You can import LDD models into Mecabricks through the File > Import function. Figure 8-22 shows a LDD project I brought into Mecabricks. Note that sometimes Mecabricks simply hasn't created those parts yet, and the model will appear with some parts missing. This is the natural result of having an application like Mecabricks that is a hobby project rather than supported by a whole corporation. Mecabricks simply haven't digitized every single brick yet.

Exporting

Currently, Mecabricks only exports in three formats: .DAE, .OBJ, and .STL. The former two are used in 3D animation programs such as Blender. The .STL format works with 3D printers, allowing you to print an entire model (Figure 8-23).

Many users are disappointed to see that they can't export to LDD or LDraw. In the latter case, it'd take a fair amount to make happen. Basically, LDraw uses a slightly different measuring system than Mecabricks. In order to

Figure 8-23 Export your model to work in Blender or print on a 3D printer.

export to LDraw or LDD, Mecabricks would have to include a conversion matrix for each part, a gigantic task for what amounts to a one-person operation.

Summary

Mecabricks is new and upcoming, with new features appearing all the time. Don't be surprised if you encounter a vastly better product in the very near future. In the meantime, learn how to share your designs in Chapter 9.

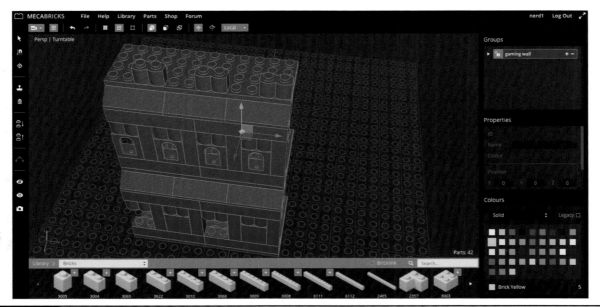

Figure 8-22 Import LDD models, though some parts won't appear.

Building Instructions

EXPERT BUILDERS LOVE to share their projects with others, and the best way to show someone how to build your model is to generate some building instructions. These are step-by-step directions generated automatically or manually and often include thumbnails of parts used in each step or inventories of all the bricks used in the model, as well as an illustration of each stage of the build process.

As I mentioned in Chapter 1, I also use building instructions to help me re-create old models that I need again. I don't have to remember how to build it; I can just look up the steps and follow them again. I have a basic LEGO Mindstorms rover I call my "minitank" (Figure 9-1). I pretty much use this as a starting point for all small robots, and I keep track of how to make it for my own memory and to share with others.

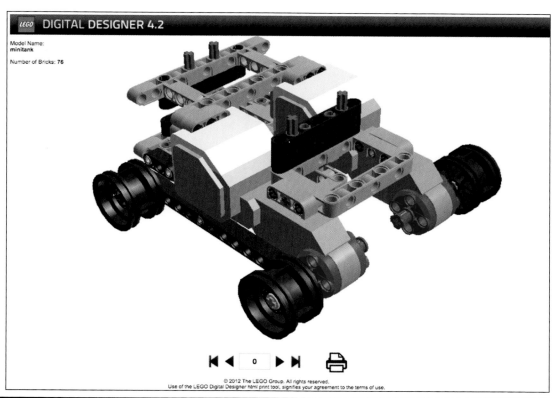

Figure 9-1 Whenever I need a small rolling vehicle for a model, I begin with my minitank.

In this chapter I'll cover the two primary tools for generating LEGO building instructions: LEGO Digital Designer's (LDD's) tool and LPub, one of the constellation of third-party tools created to go along with the LDraw Project.

Creating Instructions in LDD

LDD has a robust building instructions component, allowing you to generate a step-by-step building guide (Figure 9-2) at a moment's notice. While at first glance it appears to be nearly miraculous, there are some serious flaws. I'll unpack the good and bad aspects of the application in the following paragraphs.

Autogenerating LDD Instructions

It's almost embarrassingly easy to generate build instructions in LDD. Simply click the button on the upper right-hand corner. The button has a green brick and a yellow brick and numbers next to them. This is the Building Guide Mode, which you can also trigger by hitting F7.

Without asking for confirmation, the program builds the instructions (Figure 9-3), which can take half a minute for models with a large number of parts. The program is creating one page per step, with every piece of information you'd need to re-create that step.

There are three basic ways to use LDD's instructions. The first is through LDD, the second is by printing or making a .PDF file, and the third is to create a web page that has all the

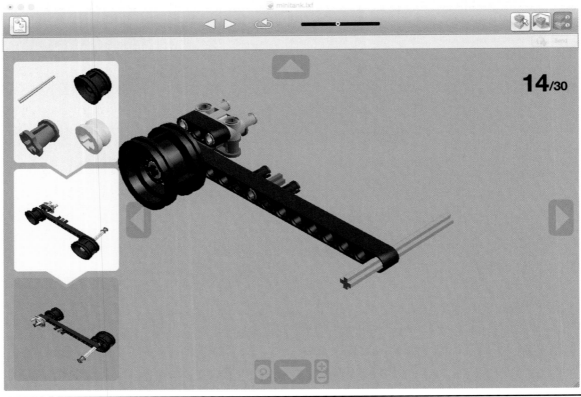

Figure 9-2   LDD's building instructions autogenerate on command.

Figure 9-3   An autogenerated instruction set.

links, thumbnails, and so forth you need to put your instructions online.

Disadvantages of the System

For a free tool, LDD's building instructions engine is top-notch. This is not to say that it's perfect, however. The following are some common complaints of LDD users:

■ **Weird part order.** Sometimes the system has a step that's silly, such as a bushing floating in space (Figure 9-4) because the cross-axle it would attach to hasn't been placed yet.

■ **LDD's autogenerated callouts seem almost random and typically not very helpful.** Wouldn't it be great if complicated substructures included a close-up zoom?

■ **Can't customize it.** The main complaint I've heard is that you can't customize LDD's building guide in any way. It's the tradeoff between autogenerating and manually generating your instructions.

■ **Large models are scaled back** (Figure 9-5) in the HTML version of the building guide. In a hypothetical step, if you add a peg to a 3-inch model, you might not be able to see where it goes.

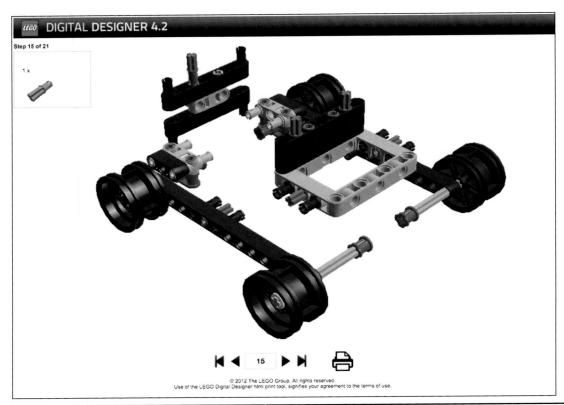

Figure 9-4 Adding bushings before wheels—just one of many odd choices LDD makes.

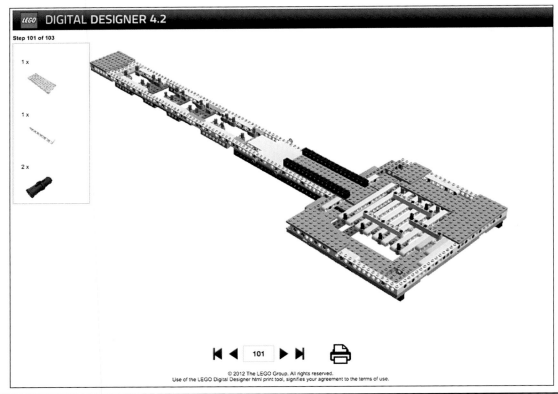

Figure 9-5 This life-sized guitar is nevertheless depicted in its entirety even though only one peg is getting added.

Steps

LDD's instructions take the form of one element per step. You can see a sample step in Figure 9-4. When you view a model's instructions in LDD's Building Guide Mode, each part is shown animated, zipping in from a margin to pop into place. I find this very useful because sometimes it's hard to get a sense of how each part fits in. You can repeat the animation as many times as it takes for you to get it.

Figure 9-6 shows how a typical page looks in LDD. The model is front and center, with navigational arrows identical to those in the Building Mode. The parts added in that stage are displayed in a window in the upper right-hand corner.

A. HTML Button

This generates a web page of your instruction set.

B. Step Selector Buttons

With one brick per step, you can make your way through very quickly.

C. Replay Part Animation

This replays the animation of the part being installed, if you missed it the first time.

D. Step Selector Slider

This slider is useful if you have an uber-complicated model and it would be inconvenient to go through the steps one at a time via the buttons.

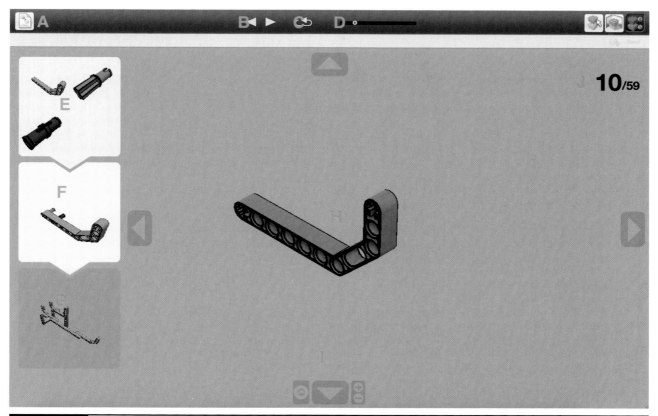

Figure 9-6 A typical LDD instruction page.

E. Parts Window

LDD shows a handful of parts that will be installed in the next few steps.

F. Parts Window

LDD shows the assembly with all the parts in it.

G. Parts Window

This pane shows the current appearance of the model.

H. Parts Preview

The main panel shows an animation of the parts' insertion.

I. Arrows

The same navigational arrows of the Building Guide Mode may be used to rotate and zoom on each and every step.

J. Step Number

This shows the current step as well as the total number of steps.

Make a Web Page

Press the "HTML" button in the upper left-hand corner of the application window to convert the building guide into a web page. You will be prompted to choose a save location for the web page. When you've done so, the system saves a single HTML page for the project, with an accompanying folder of images.

Figure 9-7 shows a sample page. Follow along with the following callouts:

A. Step number and total steps.

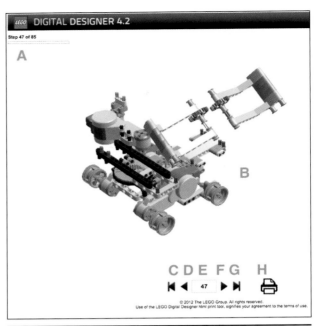

Figure 9-7   Create a web version of your building guide.

B. Beauty shot of the model, basically the completed model with the same angle as all the instructions.

C. Go to the first step.

D. Go to the previous step.

E. Go to the next step.

F. Go to the last step.

G. Print the instructions. This widget puts one instruction page on each piece of paper, printing every instruction you need to complete the model.

Differences Between the Two Versions

You should be aware that the LDD Building Guide and the HTML version of it differ in a couple of key ways:

■   In the HTML version, the steps do not animate the way they do in LDD. The HTML version features static images. Most often this looks great (Figure 9-8) but sometimes it can be a problem.

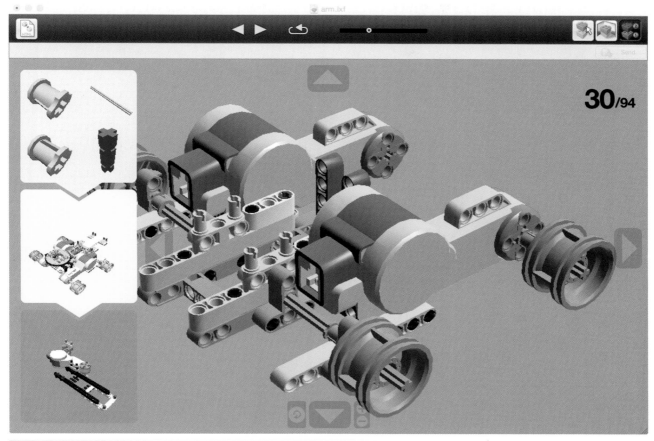

Figure 9-8   A building guide page offers a great close-up.

- In the HTML version, you lack the ability to rotate and zoom, leaving you with a zoom defaulting to the bounds of the project. As I mentioned earlier, on big projects, this does not provide a lot of detail.

Parts List

LDD also generates a list of every element used in the project, with a rendering of the part, its official LEGO part number, and the quantity used. You can see a sample parts list in Figure 9-9. This parts list includes a hidden bonus: .PNG image files of every brick used in the project, with their backgrounds knocked out. Having access to these files will prove extremely helpful if you decide to create your own custom building guide.

Printing Instructions

Once you've entered Building Guide Mode, how do you get your instructions in a different format? It turns out that there's an easy ways to accomplish this.

Once you've made a HTML version of the guide, you can print out the entire thing by clicking on the "printer" icon at the bottom of the page and then either outputting to paper or saving as a .PDF file.

LPub: Building Instructions for LDraw

There is an application called "LPub" (Figure 9-10) that is as close as it gets to an LDraw

Figure 9-9 The HTML version of the building guide includes a comprehensive parts list.

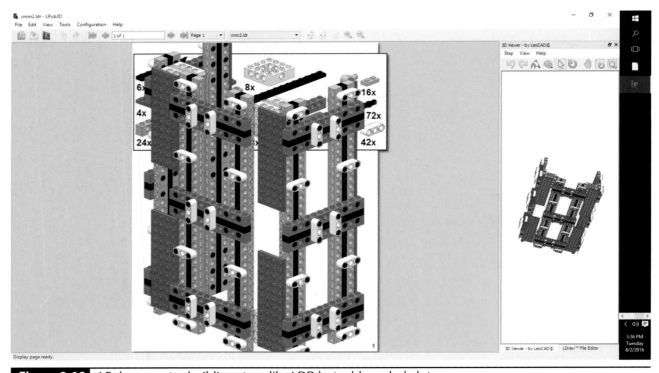

Figure 9-10 LPub generates building steps like LDD but adds a whole lot more.

application for formatting building instructions. It combines the ability to automatically generate building steps with many other features that LDD lacks.

Advantages of LPub

Fans of LPub are quick to point out some very cool aspects of the program:

- It autogenerates steps the way LDD does but gives you the option to edit every step.

- It supports subassemblies (parts of a model), and each has its own parts list and instructions, keeping large projects organized.

- It supports multiple image resolutions for online versus print usage.

- It supports Black and White Mode, where it assumes that the instructions will be printed in grayscale and adds color markers to part images as needed.

- It fades previous steps' elements so that you can more readily see the new parts.

Disadvantages of LPub

LPub users also encounter a number of challenges:

- It's a complicated program compared to LDD. If you're neck deep in LDraw, you'll likely not even notice, but those used to a more stripped-down experience and more professional-grade interface might find the hassle factor daunting.

- Not only do you need the LDraw Library installed for LPub but also the rendering software POVRay and 3D viewer LeoCAD.

Download LPub

If you downloaded one of the all-in-one installers, likely it comes with the latest version of LPub, and it's ready to go. Failing that, however, you'll have to download LPub separately (http://lpub.sourceforge.net/).

Manually Generating Instructions

Most people won't want to manually create each step of a hypothetical 100-step project. However, I have to acknowledge that the only automatic system, found in LDD, is convenient but flawed.

I happen to be one of the nerds who likes to create his own instructions (Figure 9-11), and here's why: I want to have every step make sense, not some brick-related contortion where the parts are added in a counterintuitive manner.

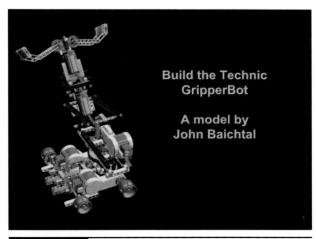

Figure 9-11   A manually generated instruction step.

Generate Step Images

1. Build the Model

I suppose that this is obvious, but it's true: first, you build the model. While you're doing so, you might want to imagine the process of building the model from nothing. However, you don't need to actually output any steps right now. Wait until the project is done!

2. Color the Latest Part

Once the project is done, it is time to work your way backwards to nothing again, one step at a time. When you're looking at the final model, you're looking at the final step of the building instructions, in which the last couple of parts were added. Find those elements, and color them some color that will stand out; I use green (Figure 9-12) or blue typically.

Whenever you design a step, try to analyze the logic behind the step. Are you maximizing the number of identical parts being added? Try to think of the most sensible option that saves you the most work.

3. Save a .LXF File for the Step

For every step, save a new .LXF file, which is LDD's native format. This is not as insane as it sounds because they are very small files: often between 5 and 50 kB. This comes out to tens of thousands of .LXF files stored on a single gigabyte of storage.

While this may seem extreme, being able to edit every single step has proven to be a great help to me over the years, and I'm sure it will help you as well. You can see a bunch of .LXF files in Figure 9-13.

4. Generate the Image

You'll need an image for your instructions, right? Simply hit CTRL+K on PCs and CMD+K on Macs, and this will create a .PNG of the step.

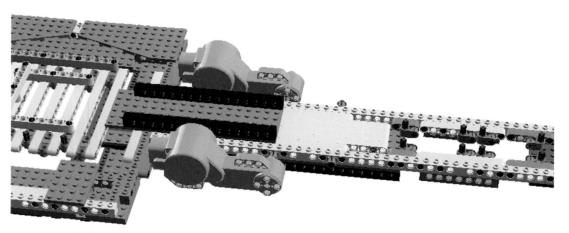

Figure 9-12 Color the latest part so that people can see what gets added.

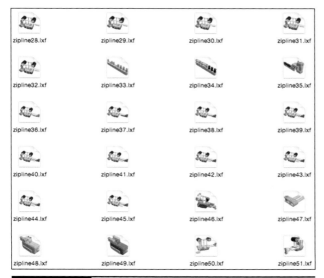

Figure 9-13 My building guides consist of a .LXF file for every step.

5. Remove the Colored Parts

After you've saved the .LXF file, delete the colored bricks, and color the elements comprising the next step. Figure 9-14 shows the penultimate step in my project.

6. Save and Continue

Repeat the process until you're done.

7. Create Ad Hoc Part Images

If you think that you might need close-ups of a particular part, create a model of that one part in LDD, zoom in to the desired level, and then create a screen shot. You can use these images for close-ups, like the one seen in Figure 9-15.

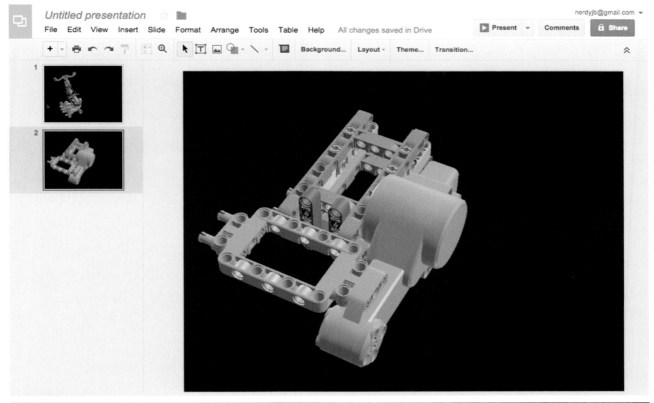

Figure 9-14 Remove the green parts, and color those that were added in that set.

Figure 9-15 This close-up of a LEGO element offers a lot more detail than the normal resolution.

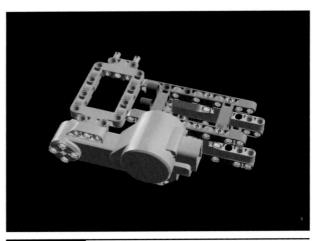

Figure 9-16 Creating a step in Google Slides.

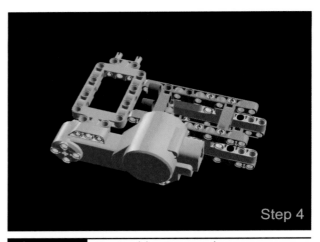

Step 4

Figure 9-17 Next, add a step number.

Work in Google Slides

Now that you have created your images, it's time to switch to Google Slides. For those of you not familiar with this resource (found at http://slides.google.com), it is a free online equivalent to Microsoft Power Point.

1. Format the Slide Show

The platform defaults to a 16:9 screen ratio, comprising the standard widescreen format, but I like working in 4:3, which is normal 8.5 × 11 landscape format. However, you can select a custom size if one of the standard ratios does not work for you.

2. Create One Page per Step

Add the new pages, and then insert the appropriate .PNG image for each step. Figure 9-16 shows the beginning of a step.

3. Add Step Numbers

Text can be easily added using Google's Slide-Numbering tool. Figure 9-17 shows a slide number added to the layout. You can also add it manually via a text box.

4. Add an Arrow

If you want to call attention to something on the slide. Use Google's Line tool to create an arrow like the one seen in Figure 9-18.

5. Create a Callout

The next step offers a bit more challenge: how to create a callout, a box off to one side (Figure 9-19) with a close-up of one part of the model. To accomplish this, take a screen shot in LDD and drop it into the slide, with a colored box underneath and an arrow pointing in the right direction.

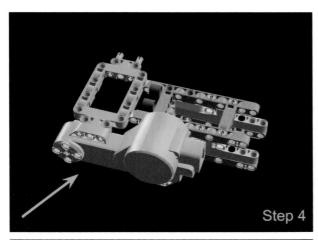

Figure 9-18 Easily add arrows to your creations.

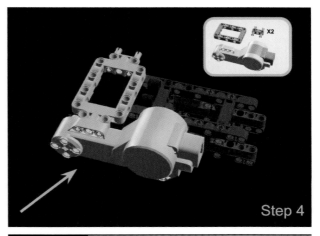

Figure 9-20 Create a fade effect with two PNGs.

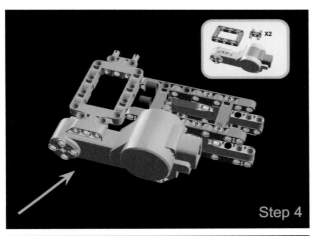

Figure 9-19 Callouts show readers an important piece of information.

the original, make another screen shot. Place the new screen shot on top of the old one, completely covering up the faded green bricks.

7. Output

From Google Slides you have a number of options for serving up the instructions to your eager readers. First, your instructions can be navigated like a slide show, either through Slides or via one of the similar applications it exports to. For instance, you can make a .PDF file of the instructions. You can also print to paper as usual, though I wouldn't recommend doing so with a black background!

6. Create a Fade

LPub boasts a neat effect in which previous steps' parts can be faded back so that they seem washed out or translucent. With a bit more work, you can accomplish the same effect in Slides (Figure 9-20). Drop in that step's .PNG as usual, and go to Image Options and reduce the image's opacity to the desired level.

Next, go back into LDD, find the .LXF file for that slide, and delete all the old parts, leaving only the recolored green bricks. Keeping the rotational angle and zoom the same as

Summary

Building instructions are important because they allow virtual builders to learn from each other. It turns out that community is important—so much so that I've devoted an entire chapter to the topic. Chapter 10 concludes this book with a look at the hobbyists who have made virtual LEGO building even cooler with their unique projects and applications.

Community

BEING ABLE TO SHARE one's creations is part of what makes virtual LEGO building so great. Online communities go without saying, but there's some even cooler projects afoot, such as participating in LDraw's part design process. There's a huge community out there—you just need to find your people!

In this chapter, I'll describe some of the online resources as well as third-party software solutions. At the end, I'll describe how to design or modify bricks so that these new elements can be used in your virtual LEGO building projects.

Websites

I've mentioned a few websites so far in this book, and here are a couple more.

Peeron

Fan site peeron.com obsessively stores LEGO-related information, featuring parts inventories (Figure 10-1) for every imaginable set. Wondering in which set that rare element can be found? If you're ever confused about what part you're looking for, look on Peeron.

Peeron also features a color guide (Figure 10-2) that gives you the lowdown on every single official LEGO color, past and present.

Bricklink

Another site useful to LEGO fans, Bricklink.com (Figure 10-3) serves as a marketplace for folks looking to buy bricks to use in their projects. You can upload a list of elements to the site and use it as a wish list to buy all the parts you need.

Lego.com

Let's not forget the mother lode. LEGO allows LEGO Digital Designer (LDD) users to upload directly to their Gallery using a dedicated button on the application (Figure 10-4). Watched over by LEGO's people, the Gallery is a safe place for kids and newcomers. And with well over half a million models on the Gallery, the quality is all over the place (go to http://ldd.us.lego.com/en-US/gallery).

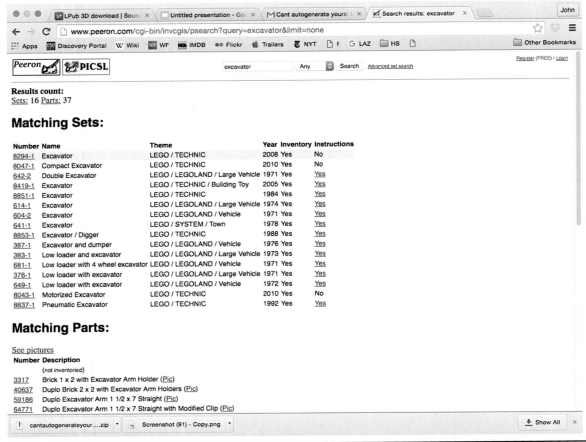

Figure 10-1 Every set and part that has the word "excavator" in it.

PearlVLGray	0	Pearl Very Light Gray	119	150	ABADAC						Belville, Bionicle
Pearlescent	47	Pearl White	83	183	F2F3F2						
PearlescentPink	4										
Pink	693	Pink	23	13	FC97AC	16	Pink				Disc. 2004
Purple	987	Purple	24	22	81007B	104	Bright violet	6B327B	63,95,0,0	2592 C	
Red	119562	Red	5	4	C91A09	21	Bright red	C4281B	0,100,100,0	032 C	
RedBrown	10624	Reddish Brown	88	70	582A12	192	Reddish Brown	694027	32,80,95,50	499 C	
RedPink	26										Clikits
RoyalBlue	219	Blue-Violet	97	112	6874CA	112	Medium Bluish Violet	6874AC	70,52,0,0	2726 C	Intro 2004, last 2005, KKII
Rust	20	Rust	27	216	B31004	216	Rust	8F4C2A	18,82,97,30	174 C	
Salmon	187	Salmon	25	12	F2705E	4	Brick Red				
SandBlue	1019	Sand Blue	55	379	6074A1	135	Sand blue	74869C	63,38,27,4	5415 C	
SandGreen	4496	Sand Green	48	378	A0BCAC	151	Sand green	789081	60,25,53,10	624 C	
SandPurple	60	Sand Purple	54	373	845E84	136	Sand violet	877C90	49,51,29,5	666 C	
SandRed	560	Sand Red	58	335	D67572	153	Sand red	957976	37,54,48,10	4995 C	
ScalaGray	406	Very Light Gray	49	503	E6E3DA	103	Light grey	C7C1B7	22,20,27,0	421 C	Disc. 2003
Smoke	1789	Trans-Black	13	40	635F52	111	Tr. Brown	BFB7B1	25,25,28,0	WarmGrey 3 C	
Tan	20819	Tan	2	19	E4CD9E	5	Brick yellow	D7C599	14,21,47,0	467 C	
Teal	1053	Dark Turquoise	39	3	008F9B	107	Bright bluish green	008F9B	100,3,47,0	327 C	
TrBlue	4277	Trans-Dark Blue	14	33	0020A0	43	Tr. Blue	7BB6E8	60,5,0,0	298 C	
TrBtGreen	122	Trans-Bright Green	108	35	D9E4A7	311	Transparent Bright Green				new 2010
TrDkOrange	1200	Trans-Orange	98	182	F08F1C	182	Tr. Bright Orange				
TrGlowInTheDark	9	Glow In Dark Trans	118	294	BDC6AD	294	Phosh. Green				
TrGreen	1974	Trans-Green	20	34	84B68D	48	Tr. Green	84B68D	60,0,60,0	360 C	
TrIceBlue	45	Trans-Very Lt Blue	113	43	C1DFF0						Clikits
TrLilac	34	Trans-Light Purple	114	236	96709F						Clikits
TrLtBlue	3066	Trans-Light Blue	15	41	AEEFEC	42	Tr. Lg blue	C1DFF0	27,0,5,0	304 C	
TrLtGreen	25										Clikits
TrLtPink	251	Trans-Pink	107	230	E4ADC8						Clikits

Figure 10-2 LEGO colors broken down by code.

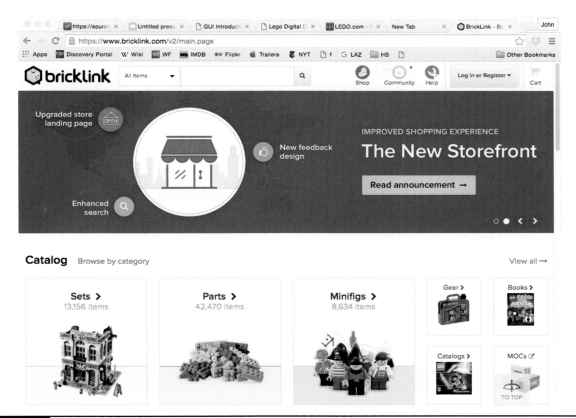

Figure 10-3   Bricklink, the marketplace for LEGO fans.

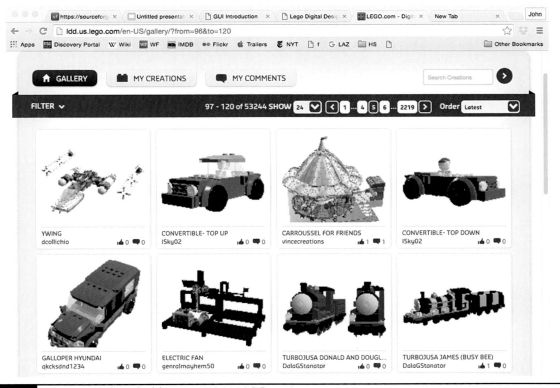

Figure 10-4   The LDD gallery holds over 500,000 LDD projects.

BrickJournal

One of the original print magazines devoted to the LEGO fan community, *BrickJournal* has been around since 2005. This journal covers every aspect of the community from train building to microbuilding or designing tiny perfect models.

BrickJournal profiles famous builders and offers tips to new ones, and it always has a presence at the annual conventions. The journal does not feature a great deal of virtual models, but maybe your art will convince the editors otherwise? Figures 10-5 and 10-6 show a couple of *BrickJournal* covers from the past few years (http:// brickjournal.com).

Software

I've already mentioned rendering software, which is software that converts the data of brick shapes into things that look like bricks on screen. I just want to mention two of these—an LDraw viewer called "LDView" and a rendering application called "POVRay."

Viewers

This vital category of LDraw apps allows you to quickly view bricks and models without the need for an entire editor.

LDView

This application (Figure 10-7) displays your LDraw model in three dimensiona (3D), allowing you to see all sides of the model. This

Figure 10-5 *BrickJournal* has been chronicling the adult LEGO fan scene since 2005.

Figure 10-6 *BrickJournal* explores a different theme every issue.

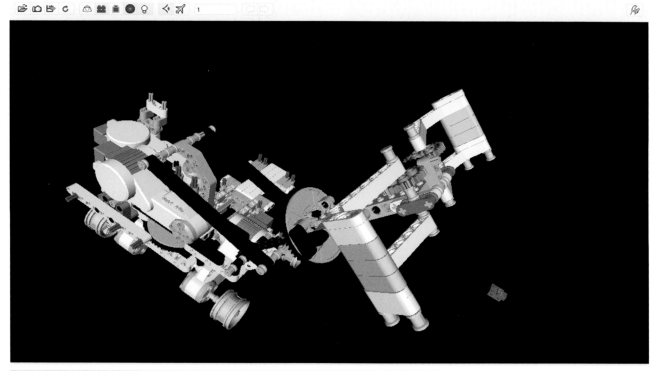

Figure 10-7 LDView lets you scroll around to see all sides of a model

is especially important because many of the LDraw editors lack the 3D scrolling we get to see in LDD and Mecabricks (http://ldview.sourceforge.net/).

Rendering

I've mentioned rendering engines before. They amp up the resolution and detail to the point where it can be hard to tell apart from a plastic model. The process can take hours to accomplish, and LEGO fans find themselves dialing down the resolution to make it go faster.

POVRay

Pretty much the most popular rendering engine in the virtual LEGO building community, POVRay makes LDraw files look really great. Figures 10-8 and 10-9 show before and after views of an LDD file exported to LDraw and

rendered with POVRay (go to www.povray.org/download/).

Creating New Parts for LDraw

One of the most exciting ways to get involved with the virtual modeling community involves actually creating your own bricks. As they stand today, neither LDD nor Mecabricks accepts custom parts, leaving only LDraw to assist. If you want an obscure part found years ago in just a couple of sets, the LDraw folks will try to re-create it. Typically, these aren't crazy, noncanonical bricks. More likely they're official LEGO parts that simply have never been digitized. Nevertheless, LDraw folks provide the tools that you can use to create your own bricks.

Figure 10-10 shows an LDraw part close up. At one point, every curve and angle in this element was hand coded by an LDraw fan.

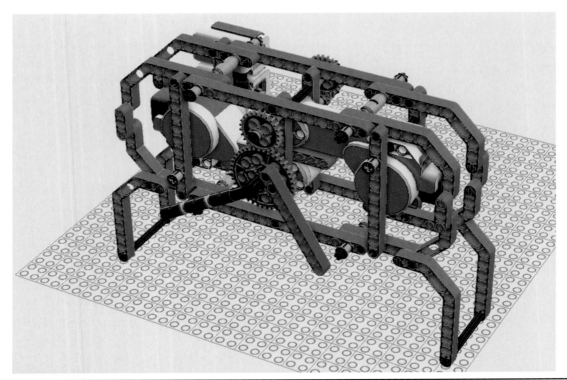

Figure 10-8 This LEGO project looks okay with a normal LDD rendering.

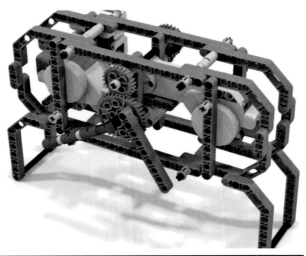

Figure 10-9 But it looks even better after being rendered!

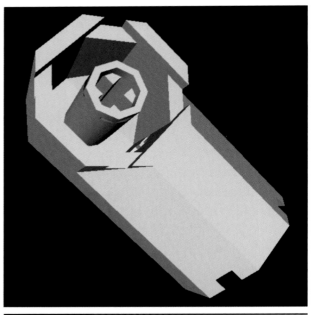

Figure 10-10 Designing bricks for LDraw helps the Library get even better.

.DAT: The LDraw File Format

The way LDraw makes and stores shapes is the .DAT format, described on the LDraw website (www.ldraw.org/article/218.html). Put simply, .DAT files consist of textiles that describe a series of shapes that are drawn between points described using X, Y, and Z coordinates. You'll see one command per line, with no limit on the number of lines in a .DAT file. Dimensions are measured in "LDraw Units" (LDUs) that measure around 1/64 inch or 0.4 millimeters.

The building program or viewer looks for a prefix beginning each line:

- 0: comment or meta command
- 1: subfile reference
- 2: line
- 3: triangle
- 4: quadrilateral
- 5: optional

If a line starts with anything other than one of these integers, everything on that line is ignored.

Subfiles

Subfiles are LEGO shapes found within the design of another LEGO shape, such as the studs on a brick. Why bother redrawing the stud every time when you only need one instance of it? Here's how a subfile reference in LDraw looks:

```
sub-file reference: 1 <color> x y z a b
c d e f g h i <file>
```

This consists of the prefix 1, the color code, and the X, Y, and Z of the part's coordinates in 3D space, with a–i describing the part's angle of orientation. The final part of the line is the filename of the subfile.

Example .DAT File

You can examine—or edit!—any .DAT file you want simply by using a text editor. The following is the listing for the classic 2 × 4 LEGO brick (Figure 10-11). It consists of the subfile 3001s01.dat, which consists of most of a 2 × 4 LEGO brick, missing only two sides, the idea being to design the bricks modularly to save work on making new bricks. In this case, the partial brick design can be added to other shapes to make more than just a 2 × 4 brick. For instance, two together could make a 4 × 4 brick. Next is the listing; note the two quadrilateral shapes (the lines starting with 4) that close out the partial brick. Figure 10-11 shows the brick in LDView.

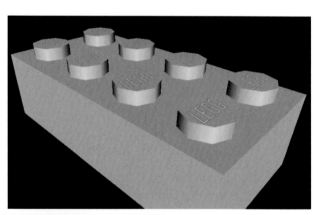

Figure 10-11   Open the file, and you'll see a classic 2 × 4 brick.

```
0 Brick 2 x 4
0 Name: 3001.dat
0 Author: James Jessiman

0 !LDRAW_ORG Part UPDATE 2004-03
0 !LICENSE Redistributable under CCAL version 2.0 : see CAreadme.txt

0 BFC CERTIFY CCW
0 !HISTORY 2002-05-07 [unknown] BFC Certification
0 !HISTORY 2002-06-11 [PTadmin] Official Update 2002-03
0 !HISTORY 2004-02-08 [Steffen] used s\3001s01.dat
0 !HISTORY 2004-09-15 [PTadmin] Official Update 2004-03
0 !HISTORY 2007-05-07 [PTadmin] Header formatted for Contributor Agreement
0 !HISTORY 2008-07-01 [PTadmin] Official Update 2008-01

1 16 0 0 0 1 0 0 0 1 0 0 0 1 s\3001s01.dat
4 16 -40 0 -20 -40 24 -20 40 24 -20 40 0 -20
4 16 40 0 20 40 24 20 -40 24 20 -40 0 20

0
```

Modify a Part

Let's do a couple of quick projects to make modifications to a brick. We're going to change one of the colors of the brick, as seen in Figure 10-12. Simply go into the 3001.dat file and change one of the lines starting with 4 16 so that it says 4 12—recall that the prefix 4 refers to a quadrilateral, and the 16 refers to the color gray. If we change the 16 to a 12, it turns one of the sides of the brick red.

For a more challenging modification, let's take off one of the studs of a brick. This means going into the 3001s01.dat file—the parent file of the 3001 you already modified—and finding where it references the subfile stud.dat, which creates the studs. Simply turn one of the line headers from 2 to 0, which comments out that stud, and you get Figure 10-13, which shows what I came up with.

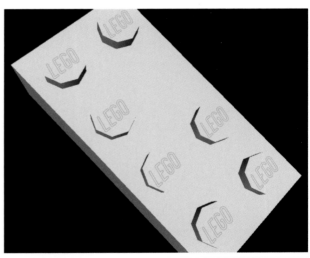

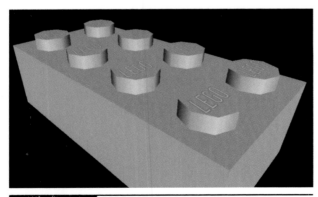

Figure 10-12 Want a brick with one red side? Design it yourself.

Figure 10-13 A seven-stud LEGO brick. Not particularly useful, but definitely cool.

Get Certified

What does it mean to get LDraw "certification" for your design? Basically, it means that a bunch of experts has looked over your design and feels that it's a faithful re-creation of the original brick. Anyone can make and use any LDraw-compatible brick, but to have it become a part of the official library takes a bit more. The following steps must be followed to get certified:

1. Design your part. LDraw focuses on files that re-create existing LEGO bricks, so you should be able to faithfully copy the design.

2. Having created an LDraw login and clicked through the contributor's agreement, go to the part submission widget and follow the directions from there (www.ldraw.org/ cgi-bin/tracker/submit.cgi).

3. Watch the Parts Tracker (www.ldraw.org/ library/tracker/; seen in Figure 10-14) to see as LDraw members certify their designs with a series of votes.

4. Once the design is approved, it becomes a Library resource and will be downloaded with all future releases of the Library.

5. While you're at it, you might want to cast a few votes yourself. At the time that I am writing this, over 600 parts still await votes.

This is not to say that you couldn't use noncertified parts. Your editor can read unofficial .DAT files just as readily as the official ones.

Figure 10-14 The LDraw Parts Tracker keeps you abreast of your submissions.

3D Printing Parts

Mecabricks exports files in the .STL format, which is compatible with most hobbyist 3D printers. This gives you the possibility of either printing bricks individually (Figure 10-15) or outputting entire models as a single piece to be printed.

Figure 10-15 3D printing bricks brings the virtual into the real world!

More intriguingly, what if you were to print LEGO bricks that do not exist and likely will not? The "Trilego brick" by Thingiverse.com user canadaduane (Figure 10-16) breaks up LEGO's usual 90-degree angles to create a series of triangular bricks.

Thingiverse user Travis Peterson designed the bike-mount LEGO plate seen in Figure 10-17. It brings your next design to the streets. The two plates bolt together onto the bicycle's top bar, allowing small models to be built on the bike itself.

A similar concept may be found in servone001's LEGO light switch plate. Great for ad hoc decorations or small skits, the light switch (Figure 10-18) adds fun to any room.

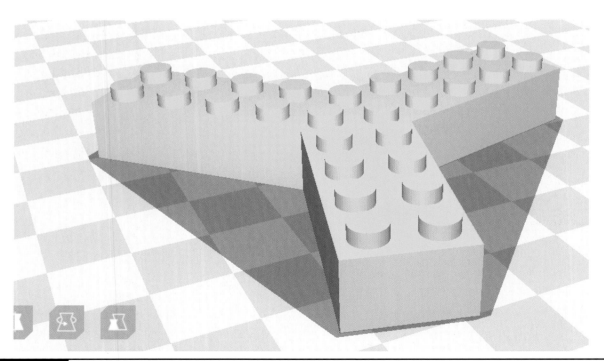

Figure 10-16 Y-shaped bricks mix up the stodgy 90-degree paradigm.

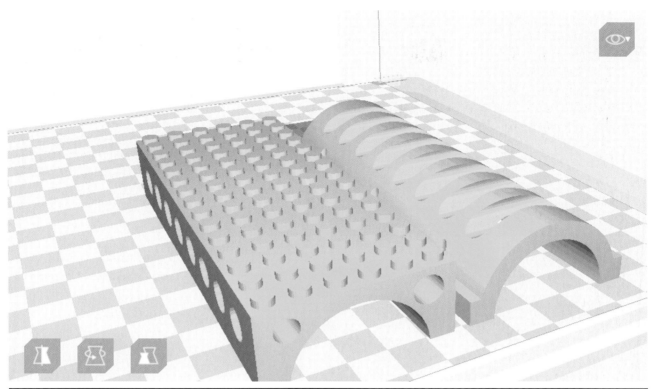

Figure 10-17   If you print this plate, your next LEGO project may be on a bicycle.

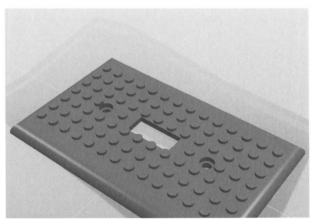

Figure 10-18   This light switch plate doubles as a prototyping area for your next creation.

Summary

With the conclusion of this chapter, we have reached the end of the book as well. I hope that your exposure to virtual LEGO building showed you how many possibilities are out there not only in terms of software but also all the related technologies, such as designing your own bricks in LDraw or printing out .STLs on a 3D printer. Good luck, and happy building!

Index

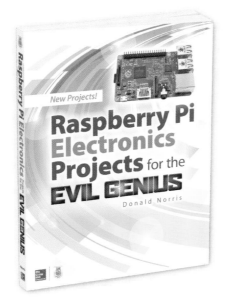
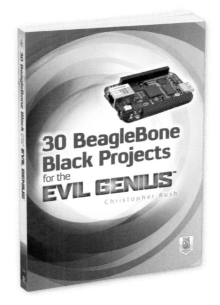
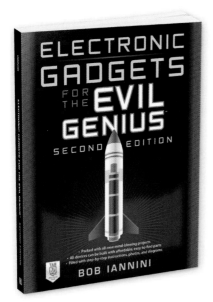
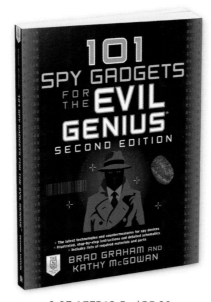
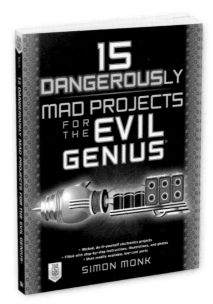